Nephrocalcinosis Calcium Antagonists and Kidney

Edited by
K.-H. Bichler and W. L. Strohmaier

With 107 Figures

Springer-Verlag
Berlin Heidelberg New York
London Paris Tokyo

Professor Dr. med. KARL-HORST BICHLER
Medical Director

Dr. med. WALTER LUDWIG STROHMAIER
Department of Urology
University of Tübingen
Calwer Str. 7
D-7400 Tübingen, FRG

ISBN 978-3-642-72859-4   ISBN 978-3-642-72857-0 (eBook)
DOI 10.1007/978-3-642-72857-0

Library of Congress Cataloging-in-Publication Data. Nephrocalcinosis, calcium antagonists, and
kidney / edited by K.-H. Bichler and W. L. Strohmaier. Includes index.
1. Kidneys – Calcification. 2. Calcium Antagonists therapeutic use. I. Bichler, K.-H. II. Strohmaier, W. L. (Walter Ludwig), 1957–. [DNLM: 1. Calcium adverse effects. 2. Calcium Channel
Blockers-therapeutic use. 3. Kidney Failure, Acute drug therapy. 4. Nephrocalcinosis drug
therapy. WJ 356 N4388]
RC916.N45 1988   616.6'106.1 dc 19   87-32317

2122/3130-543210

List of Contributors

You will find the addresses at the beginning of the respective contribution

Bachmann, S.   7
Bichler, K.-H.   '113, 126
Chaimovitz, C.   134
Deetjen, P.   34
Dirks, J. H.   43
Garthoff, B.   156
Goligorsky, M. S.   134
Hertle, L.   156
Kiryati, A.   134
Kol, R.   134
Kriz, W.   7, 113
Laberke, H. G.   67
Lach, S.   134

Lang, F.   34
Lehmann, H. D.   52
Leonetti, G.   182
Nelde, H.-J.   113
Neumayer, H. H.   164
Paulmichl, M.   34
Rapoport, J.   134
Sakai, T.   7
Strohmaier, W. L.   113, 126
Sutton, R. A. L.   105
Wagner, K.   164
Yehuda, J.   134
Zanchetti, A.   182

Preface

The effect of calcium antagonists on heart muscle and blood circulation is the reason that they have found widespread clinical application for a number of years. Less well known, in contrast, is the effect this group of substances has on the kidneys, both on kidney cells and the blood flow through the kidneys. This effect was the subject of a workshop we organized in Tübingen in June 1986. Different groups studied the effects of these substances, especially in animals, on the processing of calcium by the kidney cells and on blood flow. A possible explanation is that the calcium antagonists block the influx of calcium through special cell canals, especially the cells of the distal tubule. It is necessary to test whether there is a blockade or only a reduction in the passage of calcium.

Our understanding of the effect of calcium antagonists is in large part based on the results of morphologic, physiologic, and pharmacologic studies of calcium in the kidneys. The particular processes involved in nephrocalcinosis are special objects of study with regard to calcium antagonists. This book presents the results of experimental studies of the effect of calcium antagonists on nephrocalcinosis and acute renal failure after ischemia. In this context, the clinician is particularly interested in the use of calcium antagonists to protect against the kidney in urolithiasis, in acute renal failure and during kidney transplantation. The book is thus of interest to urologists and nephrologists as well as pharmacologists, biochemists, physiologists, and others in research.

Tübingen, October 1987

K.-H. BICHLER
W. L. STROHMAIER

Contents

Introduction

While calcium antagonists have been employed in the treatment of heart and vascular diseases for some years now, the influence this group of substances has on the kidney – that is, on the renal cell, the blood flow, and finally the functioning of the organ – has received little attention. In recent years different groups have investigated the effect of calcium antagonists on the handling of calcium by renal cells or on renal function.

As far as the renal cells are concerned, it is necessary to remember that about 50% of the filtered calcium is absorbed along the proximal tubule. The calcium uptake takes place by means of passive and active transport mechanisms. The passive transport is through paracellular shunts, and the active transport is dependent on the electrochemical gradient for sodium. Permeability plays a vital role in the proximal tubular part, whereas it is less important in the distal part and the collecting tubuli. There is much information on the cellular mechanism in the proximal tubule, yet little is known about that in the distal part. Recent experiments on madin-darby-canine kidney cells have led us to suppose that there are calcium channels which allow for the transcellular transport of calcium. To all appearances these channels can be blocked by calcium antagonists. Thus allround knowledge of the cellular handling of calcium in the kidney is necessary in order to employ calcium antagonists, in particular to achieve a protective effect. The results with regard to heart and vascular walls suggest that calcium antagonists protect against calcifying processes in the tubular cells of the kidney, in particular in the distal part.

To recognize the influence that this group of substances has on the cells handling calcium in the kidney, i.e., the proximal and distal tubular cells, it is necessary to consider that these substances produce a blockage of the calcium influx through specific channels. Thus we can start from the idea that calcium antagonists influence calcium passage in the distal part of the tubule. We can assume here that calcium antagonists influence the duration of the periods the calcium channels are open, this process obviously effecting a slowing-down of the influx of calcium through these channels. It is less probable that they block the influx of calcium entirely. Also of consequence to their effect on the tubular cells is their binding power to the cell membrane (in which the calcium channels are located). In this context it is necessary to determine whether calcium antagonists influence the function of the cell membrane.

We have to bear in mind that other factors – e.g., hemodynamics, hormonal influence, alcalosis, and the effect of vitamin D metabolites – also influence the effect on the tubular cells. All of these factors influence calcium excretion. Several

Nephrocalcinosis, Calcium Antagonists,
and Kidney
Ed. by K.-H. Bichler and W.L. Strohmaier
© Springer-Verlag Berlin Heidelberg 1988

investigations have demonstrated that renal plasma flow is increased by calcium antagonists, yet that the glomerular filtration rate remains the same. Whether hormonal influence has an effect on the renal handling of calcium is still unknown. In this context we have to consider, for example, the effect of the parathyroid hormone. Of further interest are investigations which demonstrated an increase in 1.25–D on calcium antagonists (verapamil); other investigations have, however, shown unchanged vitamin levels.

Apart from an understanding of the physiological processes and the endocrinological and pharmacological factors, we also have to take into consideration the morphological situation of those parts of the kidney which handle calcium. This is true to understand the normal condition as well as pathological alterations in connection with calcifying processes and particularly with nephrocalcinosis. This material is covered in the detailed chapter on the renal morphology. Knowledge of the extent to which processes like nephrocalcinosis or the development of calcium-containing concrements in the kidney or the upper efferent urinary tract can be influenced by calcium antagonists is of particular interest in clinical practice. Specifically, there is the influence on calcifying processes following inflammatory alterations in the area of the tubule or the interstitium or on calcifications of other origin. With regard to pathologic anatomy, there are different forms of nephrocalcinosis. Moreover, the clinician discriminates between a number of syndromes which are accompanied by nephrocalcinotic processes or are caused by them. Morphologically we discriminate between a primary (metastatic) and a secondary (dystrophic) type of nephrocalcinosis. From the anatomical point of view it is hardly possible to determine predispositions of certain parts of the kidney to develop calcifications in certain diseases. We can assume, however, that the tubular system is affected by calcifying processes much more often than other parts. In clinical practice nephrocalcinosis often becomes evident only when accompanied by an indicating symptom such as nephrolithiasis or, in the renal tubules, acidosis with hypokalemia. We are interested in the effect of calcium antagonists on stone formation and nephrocalcinotic processes. Experimental investigations of nephrocalcinosis induced by the administration of an athergenous diet showed typical calcifications in the area of the distal tubule. Electronmicroscopic examinations demonstrated considerable calcium deposits in the interstitium. Pathogenetically different explanations are probable. Noteworthy for the effect of calcium antagonists in these processes is that a considerably reduced calcium deposition results in the renal tissue during administration of calcium antagonists in an atherogenic diet. Corresponding measurements of calcium excretion as well as of the Tamm-Horsfall protein (a sour mucoprotein which is found mainly in the distal part of the tubule) and citrate content in urine indicate of that calcium antagonists have a protective effect against nephrocalcinotic alterations in the kidney. Experimental investigations of nephrocalcinotic processes have demonstrated that calcium antagonists have a protective effect on the renal parenchyma in uremic nephrocalcinosis. Less calcium was accumulated in the tubular cells of rats with verapamil and which had undergone subtotal nephrectomy than in those which had undergone nephrectomy but had not been given verapamil. Obviously, this effect is independent of the parathyroid hormone, so that a direct effect of verapamil on the tubular cells can be supposed.

Of particular interest regarding the influence of calcium antagonists are investigations on the effect of this group of substances in acute ischemic renal failure. Administration of calcium antagonists mitigated ischemically induced functional impairment of the kidney and ischemically induced calcifications. This is of clinical relevance in surgical interventions on the kidney in ischemia (e.g., surgery of bigger staghorn calculi). Last but not least, calcium antagonists seem to be relevant in transplantation medicine. By giving diltiazem in a small number of patients who had undergone renal transplantation it was possible for the function of the transplant to be improved or the nephrotoxicity of cyclosporin A reduced.

Fundamentals

Nephron- and Collecting Duct Structure in the Kidney, Rat

S. Bachmann[1], T. Sakai[2], and W. Kriz[1]

General Features

The rat kidney is a unipapillary kidney. In histologic sections the different regions of this kidney are well discernible, since the tubules as well as the blood vessels are arranged in a regular zonal pattern. The renal cortex, as a whole, is cup-shaped, with inverted margins, and surrounds the renal medulla.

The cortex consists of the cortical labyrinth and the medullary rays. The cortical labyrinth contains the renal corpuscles and the convoluted tubular segments; the medullary rays are made up by the straight tubular segments. Based on the various levels at which transitions occur between the different tubular epithelia, it is possible to divide the renal medulla into three parts: the outer medulla, which is subdivided into an outer and an inner stripe, and the inner medulla. The inner medulla forms a long papilla.

Nephrons and Collecting Duct System

Microanatomy and Segmentation

Nephrons and collecting ducts constitute the tubular component of the renal parenchyma. According to their origin from different embryonic primordia one may distinguish between the nephron (derivative of the metanephrogenetic blastema) and the collecting duct system (derivative of the ureteric bud), which are connnected by the connecting tubule. The morphogenetic origin of the connecting tubule is still a matter of debate (Kaissling and Kriz 1979; Neiss 1981).

The nephron is the structural unit of the kidney (Fig. 1). Each adult rat kidney contains roughly 30 000–35 000 nephrons (Baines and de Rouffignac 1969). The nephron begins in the cortex with the renal corpuscle. The corpuscle consists of a capillary tuft (glomerulus) which is pushed into a blind expansion of the renal tubule, Bowman's capsule. The tubular part of the nephron consists of the proximal convoluted tubule, the loop of Henle, and the distal convoluted tubule. The loop of Henle starts with the straight part of the proximal tubule (therefore also termed thick descending limb), which is followed by the thin descending limb, the

[1] Institute of Anatomy, University of Heidelberg, D-6900 Heidelberg, FRG.
[2] Department of Anatomy, Faculty of Medicine, University of Tokyo, Tokyo, Japan.

Nephrocalcinosis, Calcium Antagonists,
and Kidney
Ed. by K.-H. Bichler and W.L. Strohmaier
© Springer-Verlag Berlin Heidelberg 1988

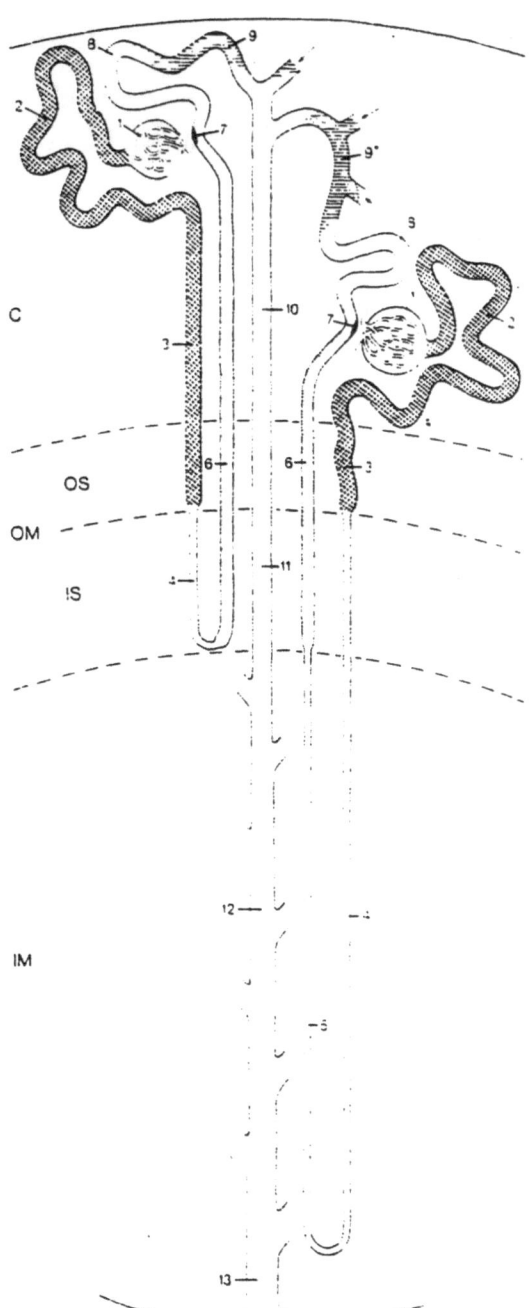

Fig. 1. Segmentation of the renal tubule. Short looped nephron on the left side, long looped nephron on the right side. *1,* renal corpuscle; *2,* proximal convoluted tubule; *3,* proximal straight tubule; *4,* distal descending thin limb of the loop of Henle; *5,* ascending thin limb; *6,* thick ascending limb; *7,* macula densa; *8,* distal convoluted tubule; *9,* connecting tubule (*9**, arcade formed by connecting tubule); *10,* cortical collecting duct; *11,* outer medullary collecting duct; *12,* inner medullary collecting duct; *13,* papillary collecting duct. *Arrows* indicate joining of other nephrons. Zonal division of the kidney: *C,* cortex; *OM,* outer medulla with outer stripe (*OS*) and inner stripe (*IS*); *IM,* inner medulla

bend (or hairpin turn) of the loop, the ascending thin limb (present in long loops only) and the thick ascending limb (also termed distal straight tubule). The nephron terminates at the end of the distal convoluted tubule (for terminology, see Table 1).

According to the location of the corpuscles in the cortex, the nephrons can be distinguished into three types: superficial, midcortical, and juxtamedullary nephrons. Juxtamedullary renal corpuscles are larger than the superficial corpuscles; the relative size ratio between superficial and juxtamedullary corpuscles

Table 1. Nomenclature of tubular segments. Divisions and subdivisions of the renal tubular are summarized; the most frequently applied terms are listed. The *serpentine arrow* indicates a gradual transition between two tubular segments. In the *right column* some frequently used abbreviations are given. (Adapted for the rat kidney, from Kriz and Kaissling 1985)

Proximal tubule	Convoluted part	P1 segment		PCT
	Straight part	P2 segment	Loop of Henle	PST
		P3 segment	Loop of Henle	
Intermediate tubule	Descending thin limb of a short loop		Loop of Henle	DTLS
	Descending thin limb of a long loop	Upper part	Loop of Henle	DTLL u.p.
		Lower part	Loop of Henle	DTLL l.p.
		Prebend segment	Loop of Henle	
	Ascending thin limb (in long loops only)		Loop of Henle	ATL
Distal tubule	Thick ascending limb ./. Distal straight tubule	Medullary straight part	Loop of Henle	MAL
		Cortical straight part incl. macula densa	Loop of Henle	CAL
		Postmacula segment	Loop of Henle	—
	Convoluted part ./. Distal convoluted tubule			DCT
Collecting duct system	Connecting tubule			CNT
	Cortical collecting duct			CCD
	Outer medullary collecting duct			OMCD
	Inner medullary collecting duct			IMCD
	Papillary collecting duct			PCD

is 1:1.15. The midcortical corpuscles tend toward the size of the superficial corpuscles.

The proximal convoluted tubule fills a major part of the cortical parenchyma. The proximal convolutions of superficial nephrons are located in the cortex corticis; portions lying underneath the renal capsule are accessible to micropuncture. The proximal convolutions of deeper nephrons correspondingly extend within deeper regions of the cortex. However. proximal convoluted tubules of midcortical nephrons eventually may ascend up to the capsule.

According to the length of the loop of Henle the nephrons of the rat kidney may be subdivided into two different types: nephrons with short loops and nephrons with long loops. The numerical ratio between short and long loops is approximately 2:1.

All short loops return within the outer medulla. The straight part of their parent proximal tubules lies within the medullary rays of the cortex and in the outer stripe of the outer medulla. Proximal straight tubules of superficial nephrons generally occupy a central position in the medullary rays. whereas those of deeper nephrons become more peripherally apposed. The transition to the descending thin limb occurs abruptly and generally marks the boundary between outer and inner stripe of the outer medulla. The thin limbs descend through the inner stripe: they are integrated into the periphery of the vascular bundles which otherwise are established by the descending and ascending vasa recta. Near the turning point of the hairpin turn the transition to the thick ascending limb occurs: this transition. abrupt as well, is constantly located near the boundary between outer and inner medulla. The length of the descending thin limb segment is therefore more or less constant. The thick ascending limb (also termed distal straight tubule) passes through the interbundle region of inner and outer stripe in the vicinity of the collecting ducts and enters the medullary rays of the cortex. It contacts its parent glomerulus at the macula densa. A short distance beyond the macula densa the thick ascending limb ends abruptly with the onset of the distal convoluted tubule (Kaissling et al. 1977). Accordingly. the thick ascending limb may be divided into a medullary and a cortical part.

Long loops reach down to the different levels of the inner medulla. The "straight" part of their proximal tubules. at least when deriving from juxtamedullary renal corpuscles, is not straight and does not lie in the medullary rays. but rather follows a tortuous course through the outer stripe of the outer medulla in the vicinity of the vascular bundles. The transition into the descending thin limbs takes place at the same level as in the short loops. The descending thin limbs pass through the inner stripe together with the thick ascending limbs in the interbundle region. They display an upper and a lower part defined by epithelial structure. The level of the gradual transition between upper and lower parts varies with the inner medullary level of the hairpin turn of a given loop. Long loops also possess an ascending thin limb segment; this segment, as defined by the character of its epithelium, begins shortly before the bend and ends at its transition to the thick ascending limb at the boundary between inner and outer medulla. together with the onset of thick ascending limbs of short loops. The thick ascending limb of the long loop passes through the inner and outer stripe close to the vascular bundles and reaches its parent glomerulus in the cortex without having entered the medullary rays. As with the short loops, the segment ends shortly beyond the macula densa.

The length of inner medullary loop segments differs markedly since the loops turn back at successsive levels. Hence. from a quantitative point of view, the number of loops decreases towards the papilla. In fact, only 1500 out of 10000 long loops reach the second half of the medulla, and only 250 loops reach the last millimeter of the papilla (Becker 1978).

The distal convoluted tubule is considerably shorter than its proximal counterpart (Kaissling and Le Hir 1984; Kriz 1967). The onset of this segment is sharply defined. Distal convoluted tubules of superficial nephrons frequently contact the renal capsule with one single convolution accessible for micropuncture; those of midcortical and juxtamedullary nephrons are located deeper in the cortex and may frequently take their course close to the medullary rays.

The collecting duct system begins with the connnecting tubule establishing the link between distal convoluted tubule and collecting duct. Microanatomically the connecting tubules of superficial and deep nephrons are different; connecting tubules of superficial (and upper midcortical) nephrons are unbranched and drain only one nephron. Those of deep nephrons fuse to form arcades before draining into a collecting duct. An arcade ascends within the cortical labyrinth and is constantly grouped around the interlobular vessels. Within the upper half of a medullary ray the arcades open into the cortical collecting duct which subsequently enters the medullary rays, mostly joining with a cortical collecting duct coming down from more superficial parts of the medullary rays. The number of nephrons drained by a single collecting duct averages six in the rat kidney (Kriz 1967); all tributaries will have joined the cortical portion of the collecting duct before it reaches the corticomedullary boundary.

The collecting ducts pass through the outer medulla (outer medullary collecting ducts) as unbranched tubules in the interbundle region. Entering the inner medulla, the collecting ducts undergo successive fusions (inner medullary collecting ducts). First fusions occur between collecting ducts deriving from the same medullary ray. The distance between fusions decreases while approaching in the papillary tip. Papillary collecting ducts open into the renal pelvis. Assuming six cortical tributaries and eight fusions of inner medullary collecting ducts, the total number of nephrons drained by a single papillary collecting duct would amount to 1536 (Jamison and Kriz 1982).

Cytologic Organization

Renal Corpuscle

The renal corpuscle consists of the glomerulus, a tuft of specialized capillaries, and of Bowman's capsule, which is a pouch-like commencement of the tubule. The narrow chalice-shaped cavity of Bowman's capsule is confluent with the tubule lumen at the urinary pole. At the vascular pole, the afferent and efferent arterioles enter and leave the glomerular tuft side by side. Through a small area between the entrance and the exit of these arterioles, the extraglomerular mesangium (see later) passes over into the mesangium proper, which forms tree-like ramifications supporting the glomerular capillary loops.

The parietal epithelium of Bowman's capsule is made up of squamous cells resting on a thick basement membrane. At the vascular pole this simple epithelium becomes the highly specialized visceral epithelium which covers the glomerular tuft (Fig. 2). At the urinary pole, the parietal epithelium transforms into the proximal tubule. The major part of the glomerular capillary wall is in contact with the urinary space and consists of three layers (Fig. 3): the visceral epithelium made

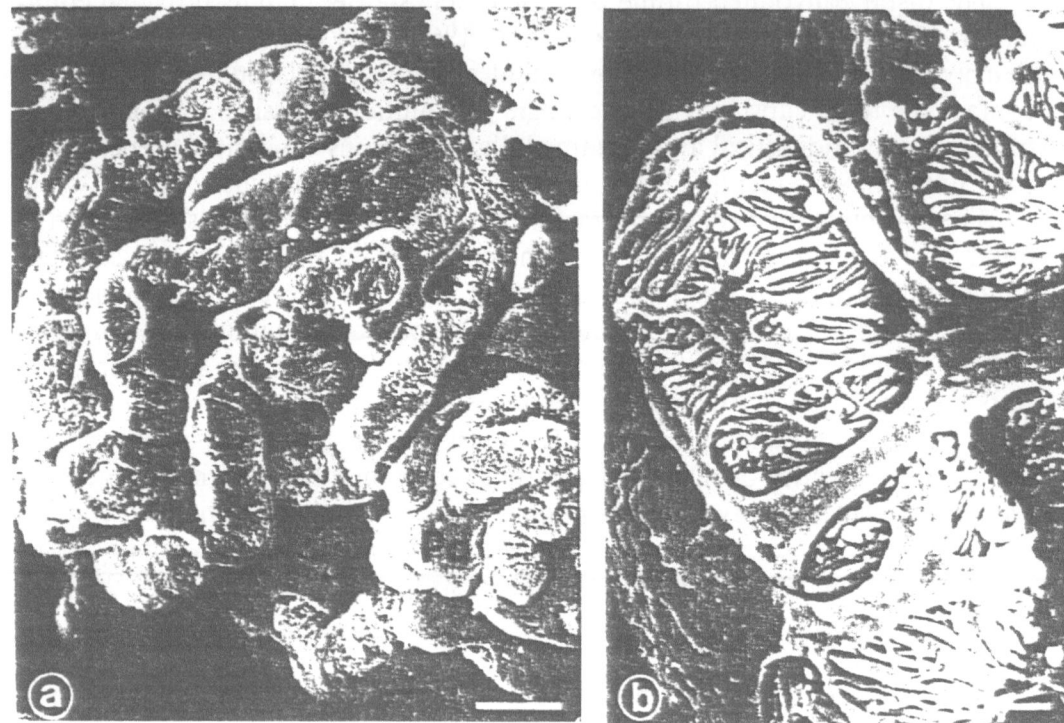

Fig. 2 a, b. The glomerulus. **a** The glomerular tuft is shown. The podoçytes (*Pd*) extend multiple cell processes surrounding the glomerular capillaries: bar = 5 μm. **b** The larger processes of the podocytes send off the foot processes. which interdigitate with those from neighboring podocytes. SEM. bar = 1 μm

up of podocytes. the basement membrane. and the capillary endothelium. In the other part of the capillary wall. the endothelium directly contacts the mesangium, which is composed of cells and a specific matrix. The numerical ratio of podocytes. endothelial cells, and mesangial cells has been estimated to be 1 : 3 : 2 (Helmchen 1980). The podocytes have a complex stellate shape (Fig. 2).

Their large cell body bulges into the urinary space and sends off long cytoplasmic processes surrounding the capillaries. The perikaryon contains a large nucleus which is deeply indented toward the side which gives rise to the ramifications. The cytoplasm on this side contains a well developed Golgi apparatus. and cisternae of rough and smooth endoplasmic reticulum. Large cytoplasmic processes give rise to many club-like terminal processes called the foot processes or pedicles. Foot processes of neighboring cells interdigitate with each other. The cell contains abundant cytoskeletal elements. Microtubules and a network of intermediate filaments are found in the perikaryon and in the primary processes, and bundles of microfibrils are observed in the apical part of foot processes (Vasmant et al. 1984).

The interlocking pattern of foot processes exhibits a very regular arrangement. The spaces between the foot processes form a long narrow meandering channel (filtration slit) with a rather constant width of 20–30 nm (Karnovsky 1979). The filtration slit is bridged by a diaphragm (slit diaphragm) of 4 nm thick-

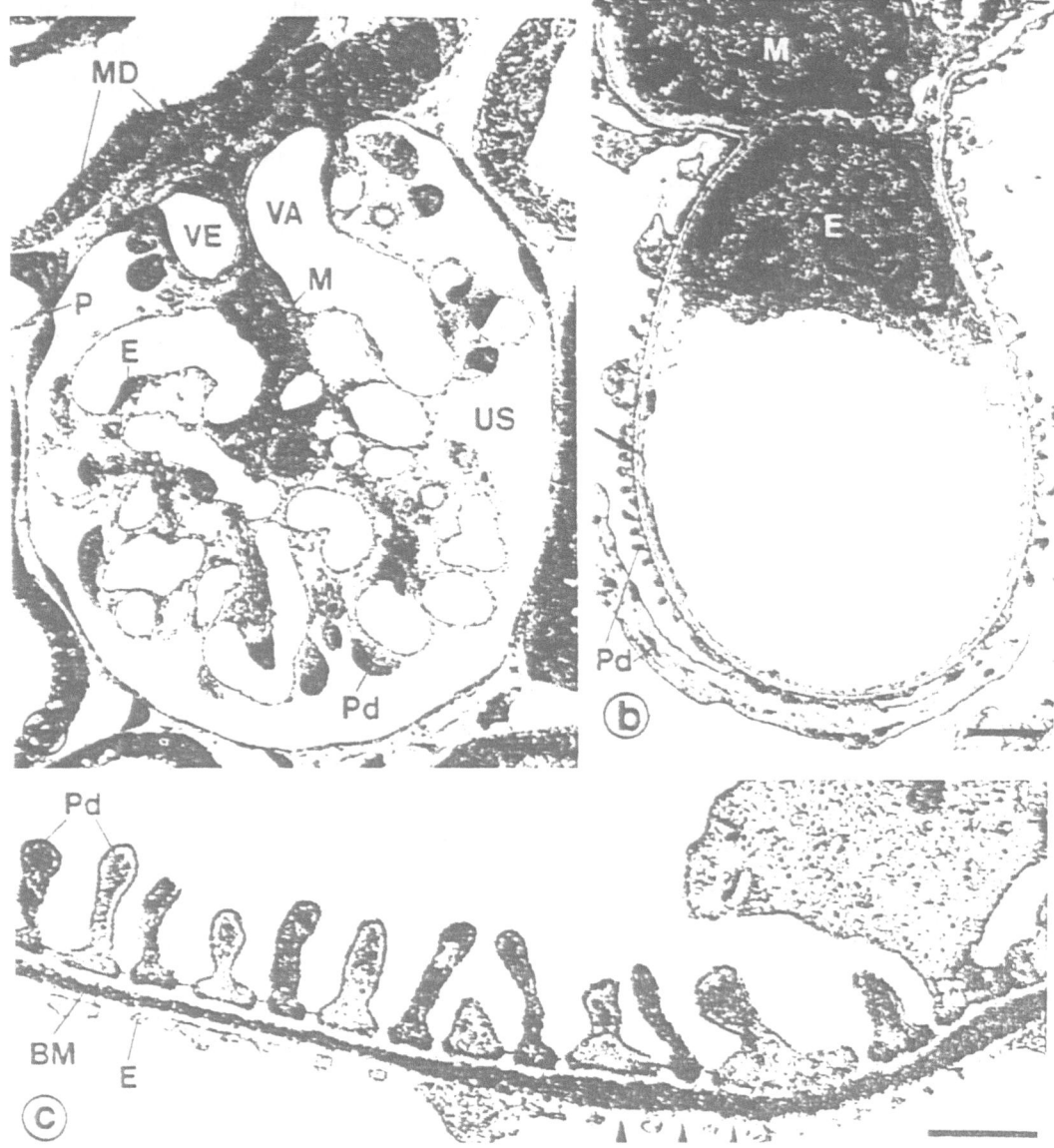

Fig. 3 a–c. The glomerulus. **a** The major components of a renal corpuscle are demonstrated. *P*, parietal epithelium; *Pd*, podocyte of the visceral epithelium; *E*, capillary endothelial cell; *M*, mesangial cell; *US*, urinary space; *VA*, afferent arteriole; *VE*; efferent arteriole. Next to the glomerular arterioles the juxta-glomerular apparatus is seen with extraglomerular mesangial cells and the macula densa (*MD*). TEM, bar = 10 μm. **b** Cross section of a glomerular capillary. An endothelial cell (*E*) is seen in contact with the mesangial cell (*M*). The basement membrane is indicated by the *arrow; Pd*, processes of podocytes. TEM, bar = 1 μm. **c** The filtration barrier. The fenestration of the endothelium (*E*) is open (*arrows*). The basement membrane (*BM*) is three layered; the filtration slits between the podocyte foot processes (*Pd*) are bridged by slit diaphragms. TEM, bar = 0.5 μm

ness. Substructure of the slit diaphragm has been demonstrated by tannic acid fixation (Rodewald and Karnovsky 1974). It consists of a central filament and regularly spaced cross bridges, which connect the central filament alternately to the cell membranes of the adjacent foot processes. This arrangement results in a zipper-like structure. The luminal cell membranes of the foot processes as well as the luminal side of the slit diaphragm are covered by a thick glycocalyx. Recently Kerjaschki and co-workers (1984) reported that podocalyxin, a 140 kd sialoprotein, is a major component of the glycocalyx of the podocytes. The abluminal cell membranes are anchored to the basement membrane by thin fibrils.

The glomerular basement membrane (GBM) of the rat has a thickness of 110–160 nm and is generally composed of three layers (Fig. 3): the lamina rara externa, the lamina densa, and the lamina rara interna. The GBM contains a network of fine fibrils which are more densely arranged in the lamina densa than in both laminae rarae. Recently, Laurie and co-workers (1984) distinguished three structural elements which compose the fibrillar network: (a) anastomosing irregular cords of 4 nm thickness, (b) straight tubular structures of 7–10 nm thickness, and (c) pairs of 3.5-nm rods. The GBM is chemically composed of (a) collagen type IV, (b) proteoglycans rich in heparan sulfate, and (c) noncollagenous glycoproteins, such as laminin and fibronectin (Farquhar 1981). Immunocytochemical studies have localized type IV collagen to the lamina densa of the GBM (Roll et al. 1980). However, there is still disagreement about the localization of the other components (Farquhar 1981). The GBM appears to be synthesized mainly by the epithelial cells (Farquhar 1981).

The mesangial cells are irregular in shape with many short cytoplasmic processes which may contain contractile filaments. A contractile response to angiotensin II and other hormones has been demonstrated in cultured mesangial cells (Foidart et al. 1980; Ausiello et al. 1980). Gap junctions connect the mesangial cells with each other, with cells of the extraglomerular mesangium, and with the granular cells of the glomerular arterioles (see below) (Taugner et al. 1978). The mesangial cells are surrounded by the mesangial matrix, which is similar in appearance to the GBM. The endothelial cells ar large and extremely flat except for the perikaryon, which usually occupies a position adjacent to the mesangium. The peripheral attenuated parts of the cell are approximately 40 nm in thickness and have many large round fenestrae (50–100 nm in diameter) without diaphragms. The fenestrations in total occupy about 20% of the peripheral capillary surface (Farquhar et al. 1961; Reeves et al. 1980). The glomerular endothelial cells are different from those of other capillaries in that they lack micropinocytotic vesicles almost completely. As revealed by cationic dyes, the endothelial cells have a thin glycocalyx (Karnovsky 1979).

The filtration barrier is highly permeable to water and electrolytes and relatively impermeable to macromolecules. The filter restricts the macromolecules on the basis of their charge, size, and configuration (Karnovsky 1979). Negatively charged cell coats of both podocytes and endothelial cells, as well as negative charges within the basement membrane GBM, are thought to be responsible for the charge selectivity of the barrier. The dense network of the basement membrane and the slit diaphragm represent the size-selective component. Molecules that have been trapped on the vascular side of the GBM are probably removed by the mesangial cells. Molecules trapped at the filtration slit may be phagocytized by the podocytes (Farquhar and Palade 1962). Apparently, the phagocytic qualities of mesangial cells and podocytes appear to be important in maintaining the functional integrity of the filter.

Proximal Tubule

Histologically, the proximal tubule is subdivided into a convoluted and a straight part. Three segments (P1. P2. and P3) are distinguished based on ultrastructural differences (Maunsbach 1966). The transition from P1 to P2 is gradual and occurs in the second half of the convoluted proximal tubule. The transition from P2 to P3 is abrupt and occurs in the first half of a straight proximal tubule.

The ultrastructural characteristics of the proximal tubule are best developed in P1 (Fig. 4). The prominent brush border is composed of long, slender, densely packed microvilli which are covered by a distinct glycocalyx (Rostgaard and Thuneberg 1972). Each microvillus contains an axial bundle of actin filaments that extends down into the apical cytoplasm (Trenchev et al. 1976). Basolaterally, the cells form a complex system of interdigitating processes. Large ridges extend from the luminal surface to the base of the cell. In the basal half of the cell, they subdivide into secondary processes that further ramify near the base of the cell. The most basal parts of the cell contain bundles of actin filaments which appear to encircle the tubule (Andrews and Bates 1984).

The junctional complex of P1 includes very shallow tight junctions which consist of only one junctional strand (Roesinger et al. 1978). This „leaky" elaboration of the tight junction corresponds well with the extremely low paracellular electrical resistance of the proximal tubules (Boulpaep 1978). Proximal tubule cells are electrotonically coupled by gap junctions.

The cytoplasm of the P1 cells contains numerous elongated mitochondria which are almost perpendicularly arranged. They are closely associated with the lateral cell membranes of the interdigitating cell processes. The Golgi apparatus is well developed and peroxisomes are numerous. The endoplasmic reticulum is mainly of the smooth surfaced type. The apical cytoplasm contains a conspicuous vacuolar apparatus which is regarded as a part of the lysosomal system (Maunsbach 1973).

Compared with P1, the structure of P2 epithelium is less complex, while the basic epithelial organization is similar. The microvilli of the brush border of P2 are reduced in height and less densely packed. The cellular interdigitation in P2 is restricted to the basal two-thirds of the cells. The vacuolar apparatus in P2 is less well developed. On the other hand, peroxisomes are generally more numerous in P2 than in P1. The tight junctions appear to be similar in both P1 and P2, but the gap junctions are less frequent in P2.

The P3 segment is structurally distinct from P1 and P2 in some respects. In the rat P3 displays the most well-developed brush border. The microvilli are the longest and most densely packed in the entire proximal tubule. In contrast, both the degree of basolateral membrane interdigitation and the association of mitochondria with the basolateral membrane are greatly reduced. The tight junctions of P3 are deeper than in P1 and P2 and consist of several junctional strands (Roesinger et al. 1978). A major function of the proximal tubule is to reabsorb organic molecules from the primary urine; the vacuolar apparatus in the apical cytoplasm is a structural correlate of the nonspecific reabsorption of polypeptides and proteins by endocytosis (Maunsbach 1973). Small molecules such as glucose and amino acids are reabsorbed by specific transport systems which are energized

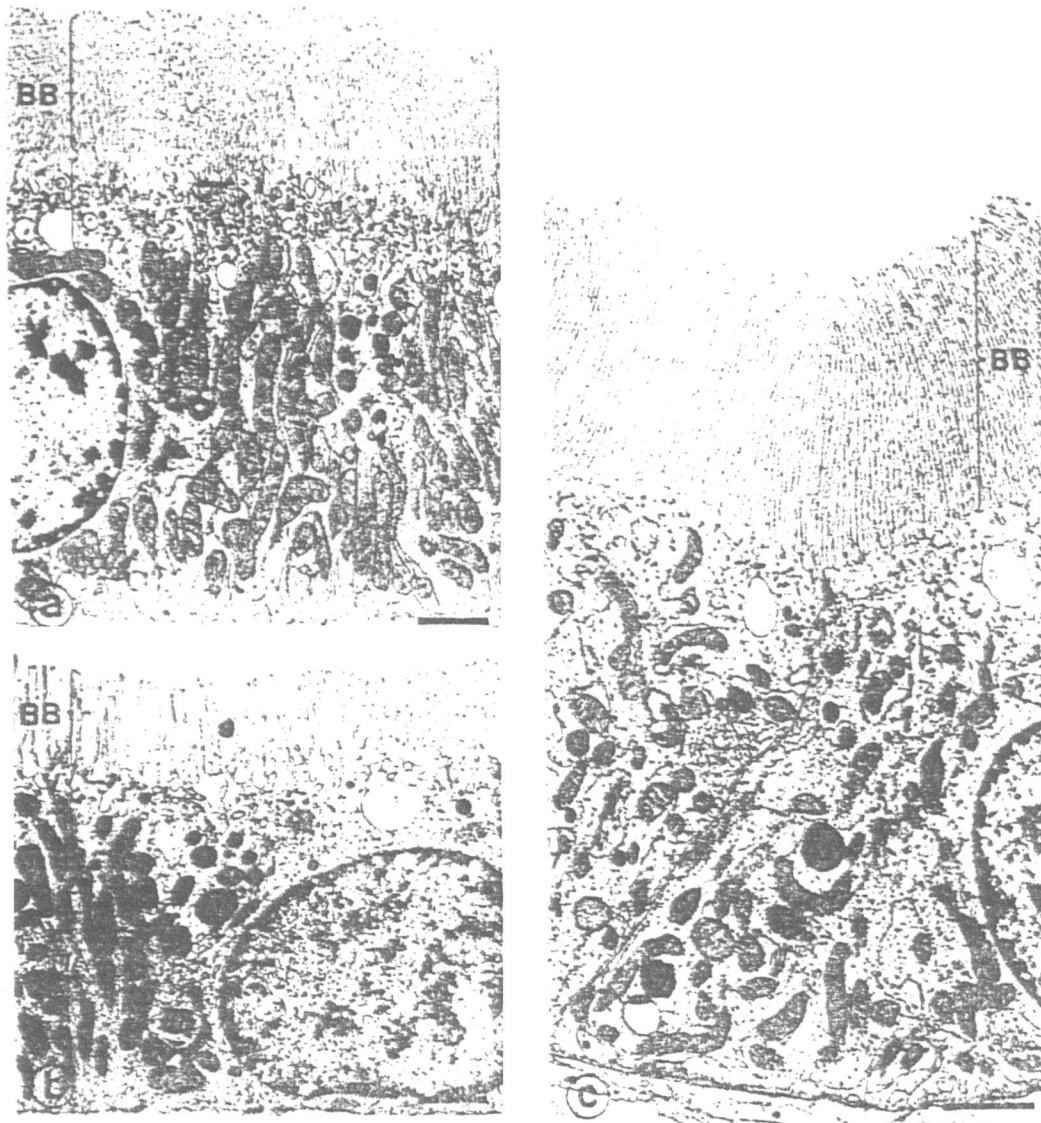

Fig. 4a–c. The proximal tubule. **a** The first segment (P1) is characterized by a high degree of cellular interdigitation: numerous mitochondria are typically associated with the basolateral cell membrane. The vacuolar apparatus (V) as well as the brush border (*BB*) are well developed. TEM, bar = 2 μm. **b** The P2 segment exhibits a less elaborate interdigitation and both the vacuolar apparatus (V) and the brush border (*BB*) are decreased. TEM, bar = 2 μm. **c** In the P3 segment the brush border (*BB*) is best developed among the proximale tubule segments. Cellular interdigitation is poorly developed. TEM, bar = 2 μm

by Na-K-APTase localized at the basolateral cell membranes (von Baeyer and Deetjen 1985; Silbernagel 1985). Glucose transport across the apical cell membrane is known to be coupled with a flux of sodium in the same direction, which is favored by an electrochemical sodium gradient maintained by the activity of Na-K-ATPase (Frömter 1979).

Thin Limbs of Henle's Loop

Four epithelial types have been distinguished ultrastructurally in the thin limbs (Bachmann and Kriz 1982; Imai et al. 1984; Kriz et al. 1972; Schwartz and Venkatachalam 1974); thin limbs of short loops possess the first type, while the three other types are found in the long loops (Fig. 5).

In the descending thin limbs of short loops (DTLS), the epithelium (type 1) is very thin (0.1–0.3 μm) in nonnuclear regions and of notably unspecialized structure (Imai et al. 1984; Kriz et al. 1972; Schwartz and Venkatachalam 1974). The cells do not interdigitate. Cell borders are bluntly apposed to each other, developing a tight junction of intermediate apical-basal depth (50 nm); desmosomes occur typically. The luminal cell membrane carries few short microvilli, mostly along the cell boundaries. Cell organelles are sparse, the majority being grouped around the nucleus.

Descending thin limbs of long loops (DTLL) possess two ultrastructurally different kinds of epithelia: the upper part of the limb is characterized by a complexly built interdigitating epithelium (type 2), while the lower part epithelium (type 3) is comparatively undifferentiated (Kriz et al. 1972; Schwartz and Venkatachalam 1974).

The upper part of the DTLL is considerably larger in luminal diameter and epithelial thickness than the DTLS (Fig. 6a). The longer the loop, the thicker is the upper part of the thin limb; therefore, the juxtamedullary nephrons possess the largest upper parts with the most complex epithelium. Characteristically this epithelium develops prominent paracellular pathways, which is evidenced by the high degree of interdigitation of the entire cells as well as by the elaboration of a lateral labyrinth formed by secondary processes of the lateral cell borders. The shallow ("leaky") tight junction is particularly lengthened due to the interdigitation of the cell borders. The luminal cell membrane is rich in microvilli. Both membranes are extremely rich in intramembrane particles (Kriz et al. 1981). The cytoplasm contains many mitochondria. Cytochemical studies revealed Na-K-ATPase (Ernst and Schreiber 1981) as well as carbonic anhydrase activity (Lönnerholm and Wistrand 1984) in this segment.

The lower part of the DTLL possesses a flat non-interdigitating epithelium (type 3) resembling that of the descending thin limb. The tight junctions are of intermediate depth (3.1 ± 0.14 strands; Schwartz et al. 1979). Infoldings of the basal plasma membrane are, however, regularly encountered, whereas intramembrane particles are inconspicuous compared with the upper part of the epithelium. The luminal membrane bears few short microvilli.

The ascending thin limb epithelium (type 4) is as flat as the lower descending part but is distinguished by an extreme degree of cellular interdigitation (Fig. 6b); the tight junctions are usually composed of a single strand. The intercellular spaces widen toward the basement membrane. The apparently leaky organization of this epithelium is partly confirmed by functional studies (Imai 1977; Marsh 1970) indicating that the ascending limbs are permeable by NaCl and urea but surprisingly almost impermeable by water. The luminal membrane bears scattered blunt microvilli; cellular organelles are inconspicuous.

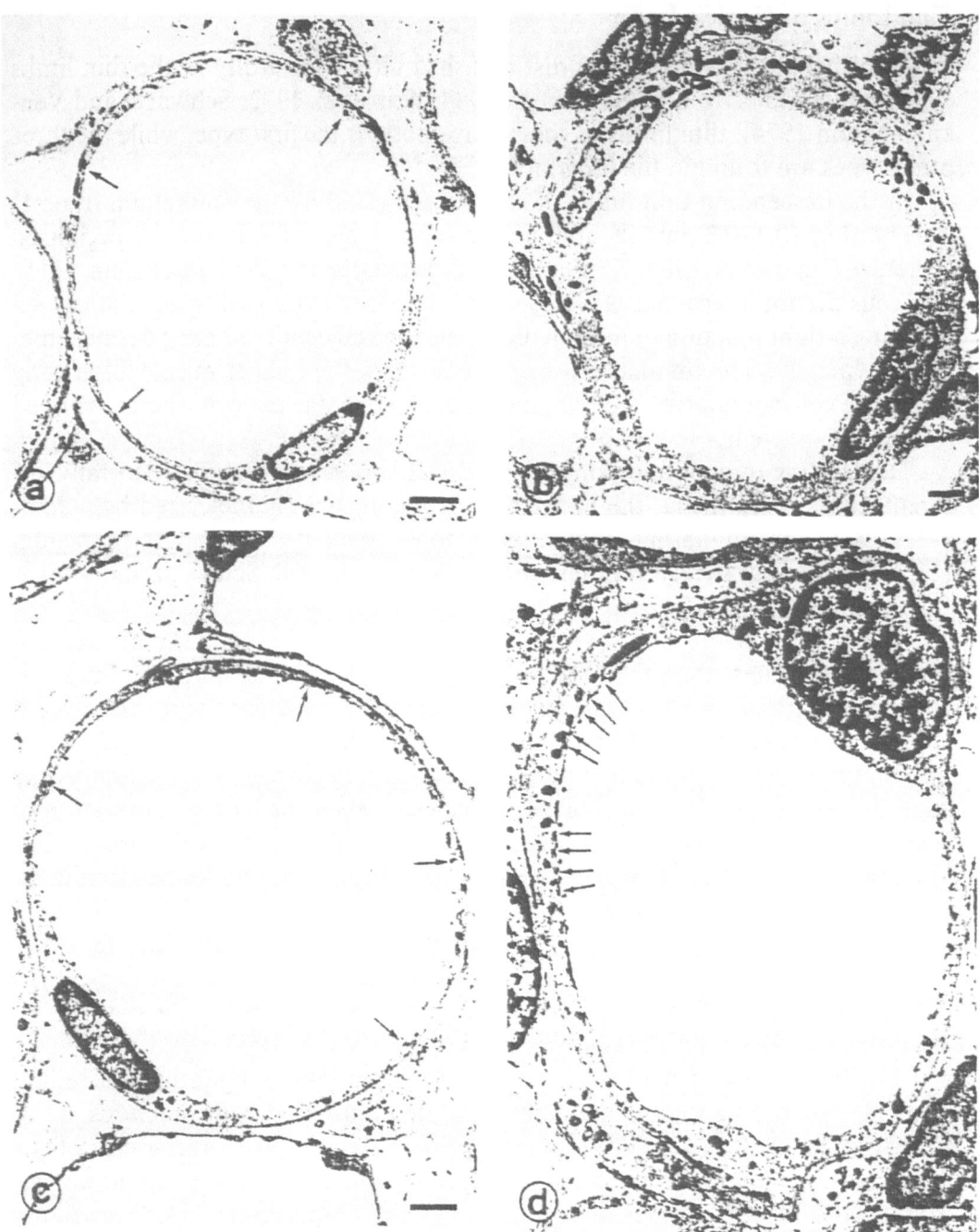

Fig. 5 a–d. Cross-sectional profiles of thin limbs of the loop of Henle. **a** Descending thin limb of a short loop in the inner stripe. The epithelium is poorly differentiated; the cytoplasm contains few organelles in the perinuclear region. Cellular junctions are indicated by *arrows*. TEM, bar = 2 μm. **b** Upper part of a descending thin limb of a long loop in the inner stripe. The epithelium is thicker and possesses numerous luminal microvilli. TEM, bar = 2 μm. **c** Lower part of a descending thin limb of a long loop in the inner medulla. The epithelium is similar to that in **a** but contains a few luminal microvilli. *Arrows* indicate cellular junctions. TEM, bar = 2 μm. **d** Ascending thin limb of a long loop in the inner medulla. The epithelium is thin and contains few cytoplasmic organelles, but the high degree of interdigitation is indicated by the numerous cellular junctions (*arrows*). TEM, bar = 2 μm

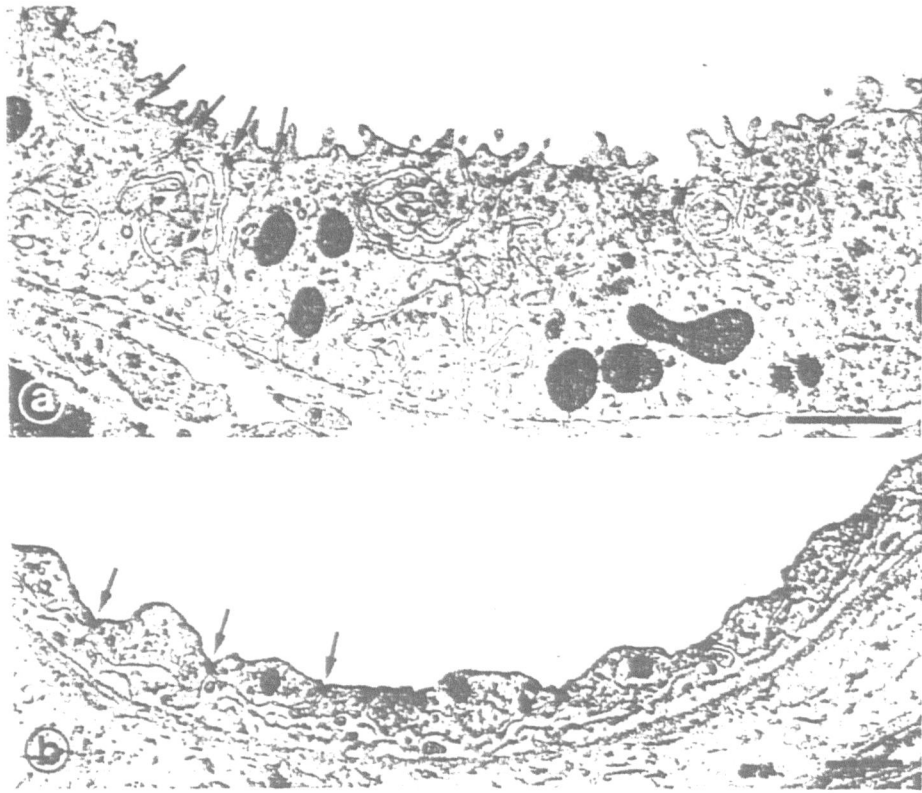

Fig. 6 a, b. Thin limb epithelia of a long loop. **a** Cross section of the epithelium of the upper part of a descending thin limb (long loop). The numerous tight junctions (*arrows*) indicate the high degree of cellular interdigitation. Interdigitation of the basolateral membrane forms a "labyrinth" of extracellular spaces throughout the entire cell body. TEM, bar = 1 μm. **b** Cross section through the epithelium of an ascending thin limb. Note the flat interdigitating cell processes. Basement membrane-like material extends into the intercellular spaces. *Arrows* indicate cellular junctions. TEM, bar = 1 μm

Thick Ascending Limb

The thick ascending limb (TAL) or straight distal tubule is lined by an epithelium that is fundamentally similar in short- and long-looped nephrons. Internephron heterogeneity, however, is evidenced by the epithelial thickness which is greater in TALs of short loops than in those of long loops (Kaissling and Kriz 1979); moreover the length of the cortical part naturally varies in that juxtamedullary nephrons are almost missing a cortical portion, whereas superficial nephrons possess the longest cortical TAL portions.

The TAL consists of one single cell type (Fig. 7). Ascending from the medulla to the cortex the epithelial structure of the TAL undergoes considerable changes. At its origin in the inner stripe, the TAL cell is tall. The cell possesses many lateral interdigitating processes which split into numerous deeply interdigitating cell processes displaying a palisade-like arrangement in the basal three-quarters of the cell. The processes contain rod-shaped mitochondria and long cisternae of the rough endoplasmic reticulum. Basally the cell processes split into small ramifications containing bundles of filaments. The apical portions of the cells have a

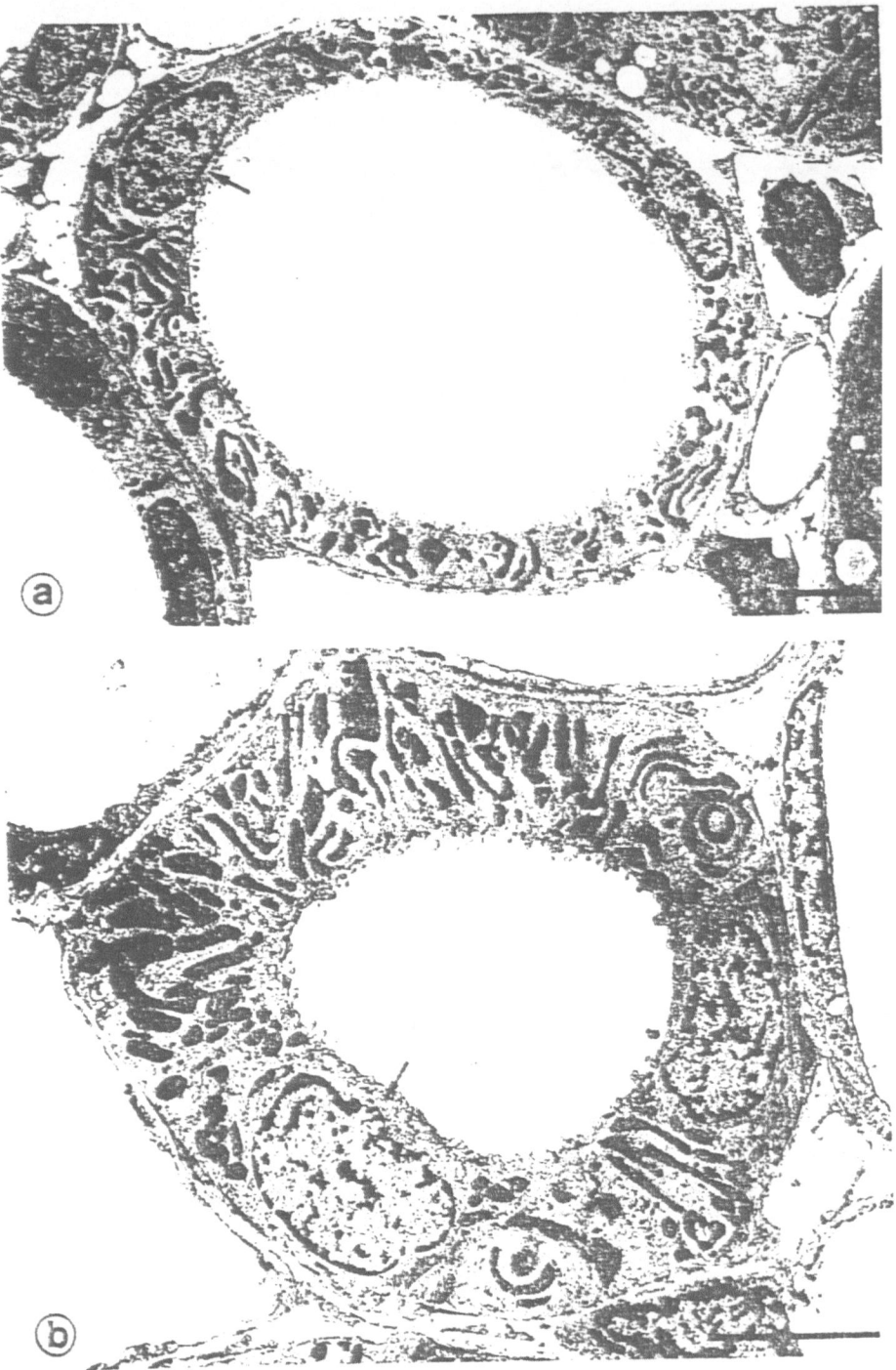

Fig. 7 a, b. Cross-sectional profiles of the thick ascending limb of the loop of Henle. **a** Cortical thick ascending limb revealing a comparatively thin epithelium with less mitochondria, which are arranged in an irregular pattern. The apical vesicular compartment contains fewer vesicles (*arrow*). TEM, bar = 5 µm. **b** Medullary thick ascending limb. The thick epithelium is stuffed with mitochondria, which are arranged perpendicularly to the basement membrane. The vesicular compartment above the nucleus is well developed (*arrow*). TEM, bar = 5 µm

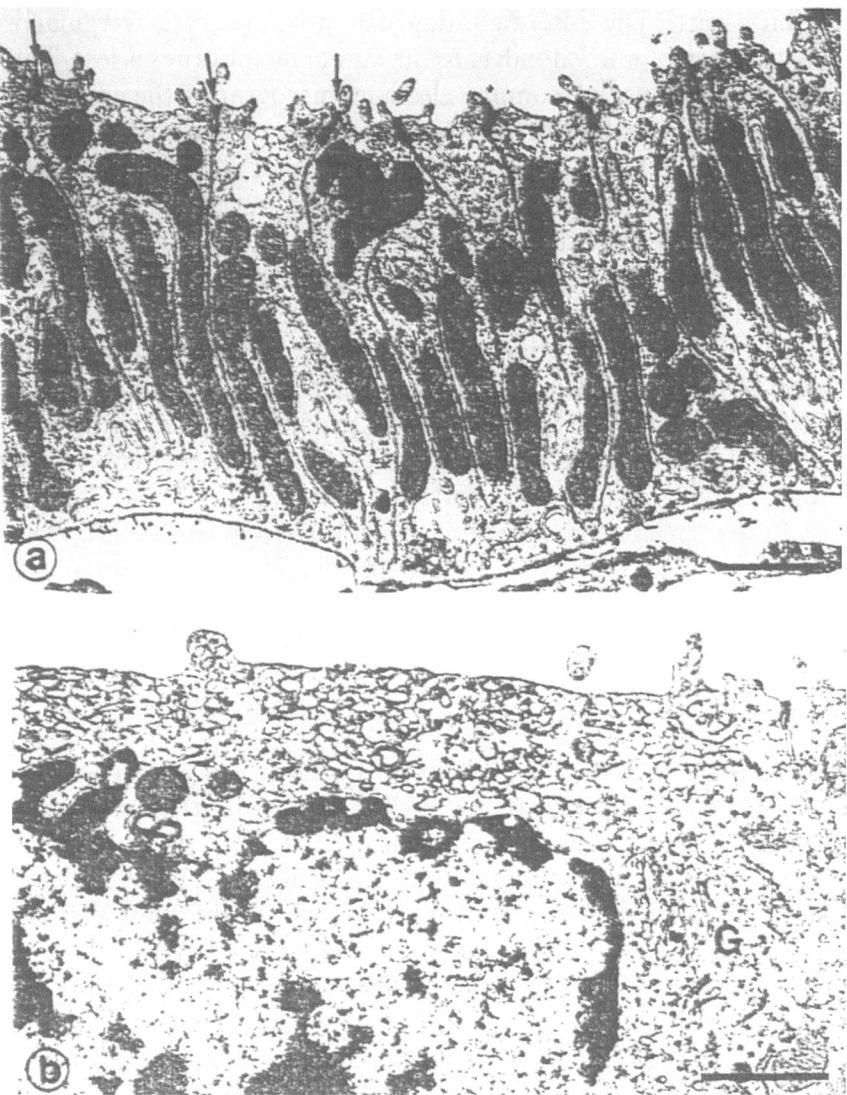

Fig. 8 a, b. Cross sections through the medullary thick ascending limb epithelial cells. **a** The cellular interdigitation reaches the apical part of the cells. Note the regular arrangement of the mitochondria, which are closely associated to the basolateral cell membranes. The apical cytoplasm regularly contains a few cytoplasmic vesicles. *Arrows* indicate cellular junctions. TEM, bar = 1 μm. **b** The vesicular compartment between the nucleus and apical cytoplasm is well developed; the multiform vesicles are membrane bound. The Golgi apparatus (*G*) regularly lies in the vicinity of the vesicular compartment. TEM, bar = 1 μm

meandering outline (Allan and Tisher 1976). Tight junctions consist of several densely arranged strands. The apical cytoplasm contains a great number of membrane-bound vesicles which constitute a proper compartment between the centrally positioned nucleus and the luminal plasma membrane (Fig. 8). Immunocytochemical methods reveal the presence of Tamm-Horsfall glycoprotein within this compartment (Bachmann et al. 1985). The Golgi apparatus is regularly found in juxtaposition to the vesicular compartment.

The TAL epithelium of the outer stripe and the cortex is strongly decreased in cell height (Fig. 8) and has, to some extent, lost the characteristic features of the deeper medullary part. The interdigitating cell processes are irregularly shaped, and the association of mitochondria to the lateral membranes is less. The lateral cell borders are tortuous; in scanning electron micrographs the epithelial surface reveals a meandering outline of the cells which indicates an elongation ot the tight junctional belt (Allen and Tisher 1976). In the outer stripe portion and especially in the cortical portion, the dimension of the apical vesicle compartment is substantially reduced. On the other hand, the amplification of the luminal membrane increases along the TAL; in the deeper inner stripe most cells have few microvilli along the cell borders, whereas in the outer stripe and, especially toward the terminal portion of the segment, most of the cells bear abundant microvilli on the entire apical surface. This axial structural heterogeneity is reflected by functional studies (Katz et al. 1979; Morel et al. 1982), which have shown a substantial decrease in ATPase activity, and differences of hormone effects in the medullary and cortical TAL portion (Kriz and Bankir 1982).

Near the end of the segment, the macula densa is situated in the TAL epithelium as a plaque of specialized cells; as a part of the juxtaglomerular apparatus its structure is described later.

Transitional Zones from the Distal Convoluted Tubule to the Collecting Duct System

In the rat kidney the three "distal" cortical tubular portions, i.e., the distal convoluted tubule (DCT), the connecting tubule (CNT), and the cortical collecting duct, do not have clear-cut delineations between each other. The transitional zones are characterized by an intermingling of the cell types characteristic of joining segments. The four cell types forming these segments are the DCT cell, the CNT cell, the principal cell (P cell), and the intercalated cell. The DCT contains exclusively DCT cells; the specific cell of the CNT is the CNT cell, and the principal cell is specific for the collecting duct. The intercalated cell is characteristically present in the CNT and in the collecting duct. Due to the intermingling of cell types in the transitional zones, intercalated cells may be found in the late DCT and principal cells already in the late CNT (Crayen and Thoenes 1978; Kaissling 1980; Kaissling and Kriz 1979; Kriz et al. 1978). In arcades even all four cell types may be found at the same level in a single CNT (Kaissling, personal communication). As a rule, however, a segment can be defined by the first appearance of a characteristic cell type (Kaissling 1982).

Distal Convoluted Tubule

The DCT cell is considerably taller than the cortical TAL cell: 18 μm vs 3–6 μm (Kaissling et al. 1977) (Fig. 9). Ultrastructurally, there are several similarities between DCT cells and TAL cells (especially those from the inner stripe portion). The basal two-thirds of the cells are split into large interdigitating processes, arranged perpendicularly to the basement membrane. Basolateral membrane amplification is higher in this segment. The apical sides of the cells abut each other bluntly; viewed from the tubular lumen, the cell borders reveal a polygonal outline. The luminal membrane carries many stubby microvilli. In cross sections the nucleus often reveals an elongated form; typically the cytoplasmic space between

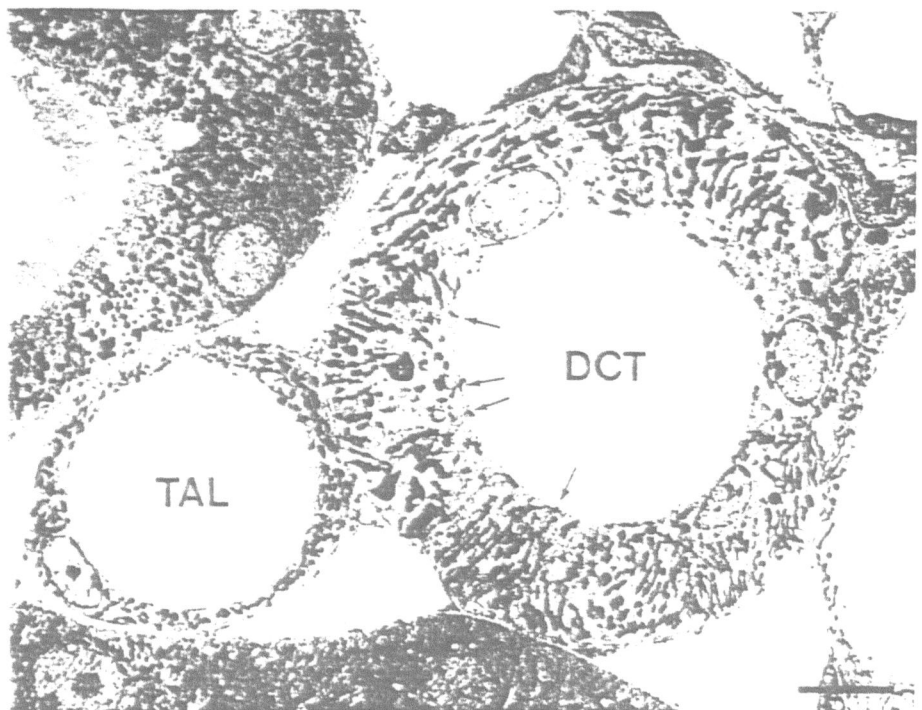

Fig. 9. Cross-sectional profiles of a distal convoluted tubule (*DCT*) and a cortical thick ascending limb of the loop of Henle (*TAL*). Note the thickness of the DCT epithelium; the nuclei of the DCT cells are generally in close attachment to the luminal plasma membrane. Few cellular junctions (*arrows*) indicate a low degree of cellular interdigitation; bar = 5 μm

nucleus and luminal membrane is narrow and contains a number of membrane-bound vesicles but rarely any mitochondria. The rod-shaped mitochondria of the interdigitating cell processes are intimately apposed to the basolateral cell membrane. This is reflected by the high Na-K-ATPase activity reported in this segment (Ernst 1975; Garg et al. 1981; Le Hir et al. 1982). Adaptation to changes in the electrolyte intake can cause significant changes in area of the basolateral cell membrane (Kaissling and Le Hir 1985), as well as in cellular height and length of the entire tubular segment (Kaissling et al. 1985).

Connecting Tubule

The two different types of CNT, the superficial CNT and the arcade-forming CNT, are cytologically equal. In both, the epithelium contains CNT cells and intercalated cells (Stanton et al. 1981). The CNT cells are tall and polygonal (Fig. 10a). Their height is comparable to that of DCT cells, but the luminal membrane above a nucleus usually protrudes into the tubular lumen, whereas the DCT epithelium does not have an elevated surface. The nucleus is centrally positioned. Like most nephron cell types, the CNT cell reveals an amplification of the basolateral plasma membrane. In CNT cells, however, the lateral membrane is barely amplified, and the cells abut each other bluntly. In contrast, the basal cell membranes have extensive infoldings, reaching almost to the luminal cell membrane. Elongated mitochondria are often located within these infoldings; however, espe-

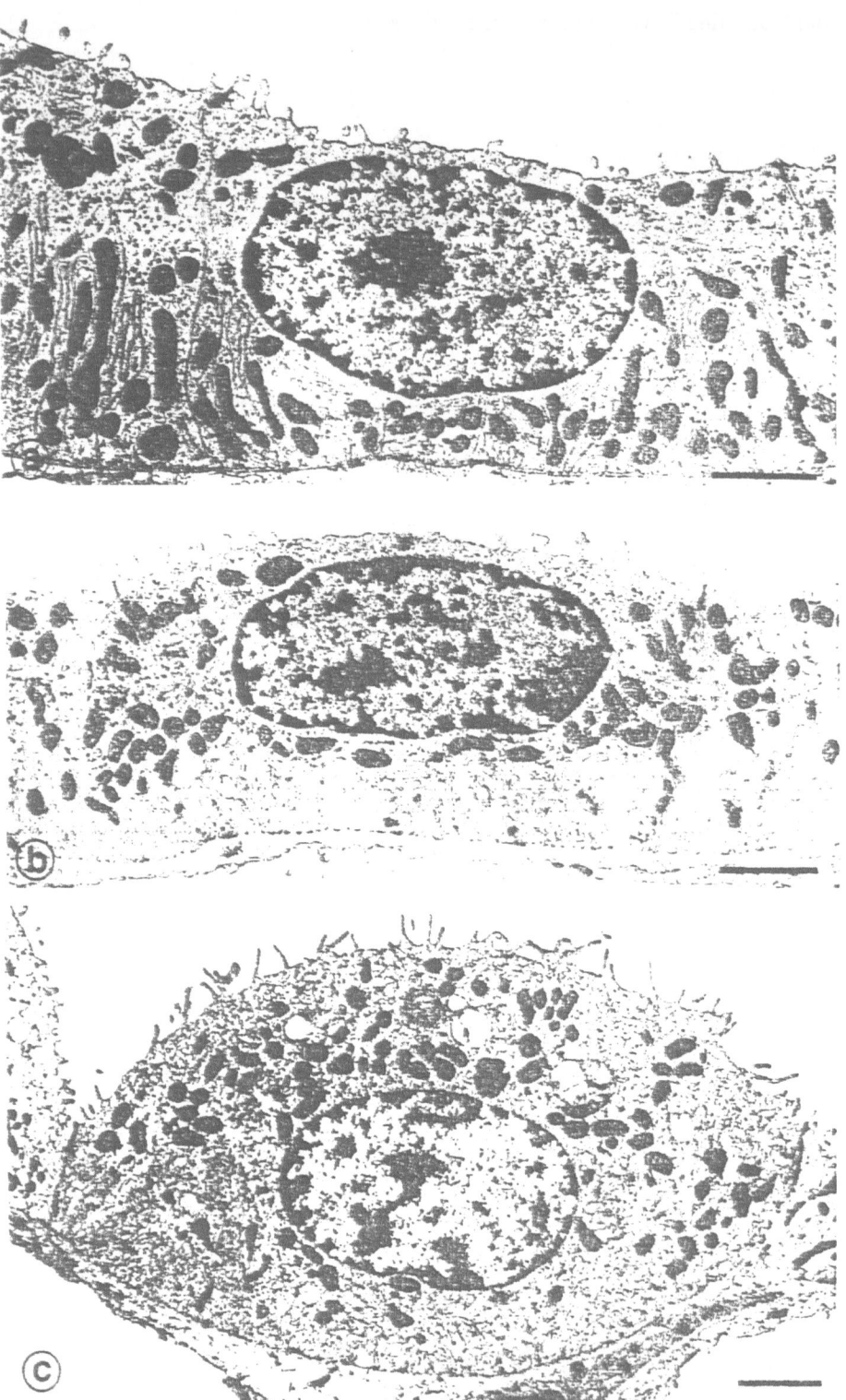

Fig. 10a–c. Epithelial cross sections of the collecting duct system. **a** Connecting tubule. The cells are not interdigitated, but the basal cell membrane displays extensive infoldings reaching far upward into the cytoplasm. The infoldings are in close relationship to basal mitochondria. TEM, bar = 2 μm. **b** Cortical collecting duct. Infoldings of the basal plasma membrane are more tightly and regularly arranged but remain restricted to the basal one-third of the total cell height. Mitochondria are found only above this basal labyrinth. TEM, bar = 2 μm. **c** Intercalated cell of an outer medullary collecting duct. The luminal cell membrane is in the expanded state with numerous microvilli and microfolds and a large cell surface bulging into the tubular lumen. The cytoplasm contains membrane-bound apical vesicles and numerous mitochondria which are apically located. The basal plasma membrane reveals meandering infoldings. TEM, bar = 2 μm

cially near the base of the cell, some infoldings are found devoid of mitochondria. The luminal cell membrane carries few distinct microvilli. Changes in the electrolyte intake have shown that in potassium-adapted animals the basolateral membrane surface density is increased by 45% (Stanton et al. 1981), and concomitant changes in Na-K-ATPase in rabbit CNT cells have been shown by Le Hir et al. (1982). The data suggest that the CNT cell is a cell type with distinct functional and structural characteristics, which clearly differ from those of DCT and principal cells.

Collecting Duct

The collecting duct epithelium of the cortex and outer medulla contains, like the CNT, two cell types. The ratio between principal and intercalated cells increases from cortex to medulla, and after the beginning of the inner medulla the principal cell becomes the only cell type forming the collecting duct epithelium.

Principal Cells. The ultrastructural organization of the principal cell is similar to that of the CNT cell (Figs. 10 b and 11). Its height is less than in the CNT cell, but it has a similar polygonal outline. The principal cells have basal membrane infoldings which are concentrated at the base of the cell, and are free of major cell organelles, such as mitochondria. The degree of amplification is smaller than

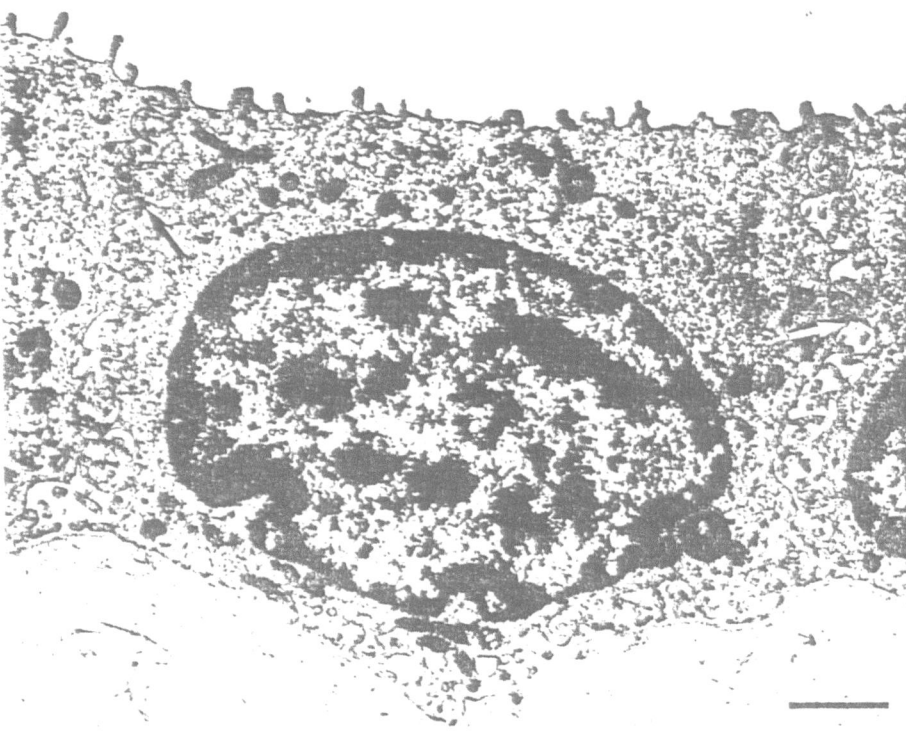

Fig. 11. Inner medullary collecting duct cell. The shape of the cell is cuboidal; lateral cell membrane amplification is evident by lateral microvilli or microfolds protruding into the lateral intercellular space. Note the desmosomal cell connections (*arrows*). To the same degree, the basal cell membrane is amplified by infoldings. The cytoplasm contains only a few mitochondria. TEM, bar = 1 μm

in CNT cells (Stanton et al. 1981; Wade et al. 1979). Unlike CNT cells, the lateral principal cell borders are interlocked with lateral microvilli and are often connected by desmosomes. Between the villi there is a significant lateral intercellular space which is not in communication with the "extracellular space" of the basal infoldings. The junction consists of several anastomosing strands forming a complex network (Pricam et al. 1974). The lateral space was found to be wide under the influence of vasopressin (up to 78% increase in width) and narrow without it (Grantham 1971; Grantham and Burg 1966; Kirk et al. 1984a, b). The apical membrane carries a few microvilli. The cytoplasm contains lysosomes to a varying extent, lipid droplets, and apical microtubuli and microfilaments (Kaissling 1982). Under conditions of increased water transport, the slow formation of large cytoplasmic vacuoles was observed in rabbit cortical collecting duct (Kirk et al. 1984b). Increases in the basal foldings, which are believed to be the effectors of sodium and potassium transport, are usually accompanied by increases in tubular Na-K-ATPase (Garg et al. 1981). They have been observed after uninephrectomy (Scherzer et al. 1985; Zalups et al. 1985), low sodium, high potassium intake, and high aldosterone levels (Stanton et al. 1981, 1985).

Axial changes of the collecting duct epithelium include a reduction of the basal membrane amplification, which is almost total in the inner medulla. In contrast, the cell size increases from cortex to medulla and is largest in the papillary duct. The epithelium near the papillary tip is cuboidal to low columnar in shape.

Intercalated Cells. The intercalated cells are interspersed in epithelia consisting of other cells (Fig. 10c) and are never seen in direct contact with one another (Fig. 12). The distribution of these cells in the final distal convoluted tubule is sporadic; in the connecting tubule they amount to 24% of all cells, in the cortical collecting duct to 32% (Stanton et al. 1981), and they then decrease in number toward the inner medulla. The intercalated cells are often called "dark" cells because the cytoplasm is frequently darkly stained. Only a few ultrastructural criteria may be used reliably for identification of intercalated cells since their struc-

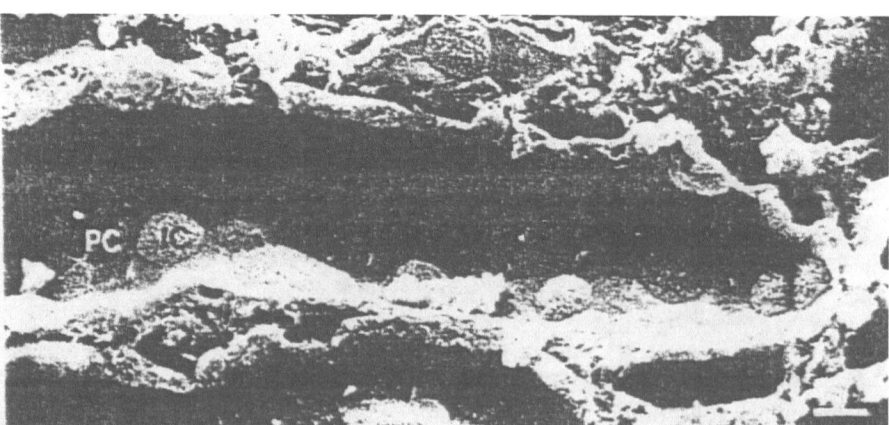

Fig. 12. Longitudinally opened cortical collecting duct. The intercalated cells (*IC*), revealing an enhanced luminal cell membrane amplification, are interspersed between principal cells (*PC*). SEM, bar = 5 μm

tural features are known to vary considerably (Hansen et al. 1980; Kaissling and Le Hir 1984; Madsen and Tisher 1983, 1984; Rastegar et al. 1980; Stanton et al. 1981; Stetson et al. 1980). The most constant features are areas densely filled with mitochondria or smooth endoplasmic reticulum, a large Golgi apparatus, and a basally located nucleus. The cell outline is more round than polygonal; infoldings of the basal membrane occur similarly as in principal cells but react differently in the physiologic experiment (Ernst 1975; Stanton et al. 1981).

Cytochemically, the intercalated cells have been shown to contain large amounts of carbonic anhydrase C (Lönnerholm and Wistrand 1984). The apical cell pole is subject to great structural alterations (Bachmann et al. 1983; Kaissling and Kriz 1979; Stetson et al. 1980). The apical cytoplasm contains a "membrane compartment" (Madsen and Tisher 1983) with small tubular and spherical structures, coated vesicles, and invaginated vesicles bearing a coat of rod-shaped particles the same as those on the cytoplasmic face of the luminal cell membrane (Stetson et al. 1980). Intercalated cells with a small surface and few microvilli possess a large membrane compartment; the cytoplasm appears to be apically constricted along the tight junctional belt of these cells. In another manifestation, however, the luminal cell pole is widely exposed and protrudes into the lumen with extensive microvilli and microplicae. The apical cytoplasm of these cells contains few vesicles (Hansen et al. 1980; Stetson et al. 1980). All intermediates between these features may be found in normal animals, but both extremes of structural appearance can be synchronized in the physiologic experiment described (Evan et al. 1980; Stetson et al. 1980). It is suggested that via a membrane "shuttle system," membrane area carrying proton pumps may either be stored intracytoplasmically or be transferred to the luminal plasma membrane (Madsen and Tisher 1983, 1984; Schwartz and Al-Awqati 1985). Apparently, different configurational changes in features of the intercalated cells reflect functionally different roles in acid-base regulation (Al-Awqati and Schwartz 1985; Verlander et al. 1985). Along the distal nephron, function of intercalated cells is believed to undergo considerable axial changes.

The Juxtaglomerular Apparatus

The juxtaglomerular apparatus is constituted by three basic elements: (a) the macula densa of the distal nephron, (b) the renin-producing granular cells of the afferent (and sometimes also the efferent) arteriole, and (c) the extraglomerular mesangium, also referred to as Goormaghtigh cells or "lacis" cells (Barajas 1970, 1981; Goormaghtigh 1939). Because of their intimate spatial as well as functional relationship, the glomerular mesangium and the unspecialized vascular smooth muscle cells of the afferent and efferent arterioles may be regarded as part of the juxtaglomerular apparatus (Taugner et al. 1978).

The macula densa is a specialized cell plaque of the thick ascending limb of Henle's loop (vide supra) adjacent to the hilus of the glomerulus (Fig. 13a). The base of the cell plaque consistently faces the extraglomerular mesangium, and, to a varying extent, touches portions of the vas afferens and vas efferens (Barajas 1970, 1971; Barajas and Latta 1963; Christensen and Bohle 1978; Christensen et al. 1979). The cells of the macula densa are clearly delineated from the surrounding cells of the TAL (Kaissling and Kriz 1979, 1982; Sikri and Foster 1981). As a whole, the macula densa protrudes into the tubular lumen: the luminal membrane bears numerous slender microvilli. In contrast to the adjacent cells of the TAL, the macula densa cells lack the Tamm-Horsfall protein (Hoyer et al. 1979). They are taller than the TAL cells, but are connected by similar tight junctions. Unlike TAL cells, basolateral interdigitation is almost lacking, and the lateral intercellular spaces extend straight from the tight junctions to the base of the epithelium. Slender microvilli and microplicae extend to the lateral space and often contact those from neighboring cells with desmosomes. Regarding functional adaptation, the lateral intercellular space is normally wide but may be found fully collapsed in furosemide and mannitol diuresis (Kaissling and Kriz 1982); on the other hand, lowering the tubular osmolarity was reported to open the lateral spaces within a few minutes (Kirk et al. 1985).

The granular cells are found mainly in the terminal portion of the afferent arteriole (Fig. 13 b), but occasionally also in the initial portion of the efferent arteriole (Barajas 1981; Bucher and Kaissling 1973). Clusters of cells are typically grouped together in the wall of the arteriole, replacing normal smooth muscle cells. Under specific stimuli, smooth muscle cells may convert into additional granular cells (Peter et al. 1974). Processes of the granular cells typically contact surrounding granular cells, cells of the extraglomerular mesangium, and smooth muscle cells, and gap junctions are frequently developed. Like smooth muscle cells, the granular cells are contacted by foot processes of vascular endothelial cells which penetrate the subendothelial basement membrane. The specific electron dense cytoplasmic granules are membrane bound and of irregular size and shape (Barajas 1966; Bucher and Kaissling 1973; Latta and Maunsbach 1962). Immunocytochemically, they have been shown to contain renin (Taugner et al. 1984). The vesicle contents are probably released via exocytosis toward the extracellular matrix between the Goormaghtigh cells.

The extraglomerular mesangial cells are situated between the two glomerular arterioles and the macula densa (Fig. 13 a) revealing a cone-shaped formation with the base at the macula densa and the apex blending with the "intra"-glomerular mesangium (Gorgas 1978). Ultrastructurally there is no clear distinction between extra- and intraglomerular mesangial cells (Bucher and Kaissling 1973; Rouiller and Orci 1971). Both are characterized by extensive ramifications containing myofibrils and by a scanty cytoplasm (Barajas 1981; Bucher and Kaissling 1973; Gorgas 1978); cytoplasmic organelles are generally scarce. The extraglomerular mesangial cells are separated from each other by a basement membrane-like ground substance but abut each other in places by gap junctions (Gorgas 1978; Taugner et al. 1978). Hence, ultrastructural features like the development of gap junctions and myofibrils present in all three cell types of the juxtaglomerular apparatus may be regarded as evidence for their common development from smooth muscle cells. In the light of the central position of the juxtaglomerular apparatus and the consistent spatial relationship to the macula densa on the one side and to all other components of the juxtaglomerular apparatus on the other side, it may be well considered that the extraglomerular mesangium performs the receptor role in the glomerular feedback mechanisms (Forssmann and Taugner 1977; Gorgas 1978; Schnabel and Kriz 1984; Taugner et al. 1978). Generally, the

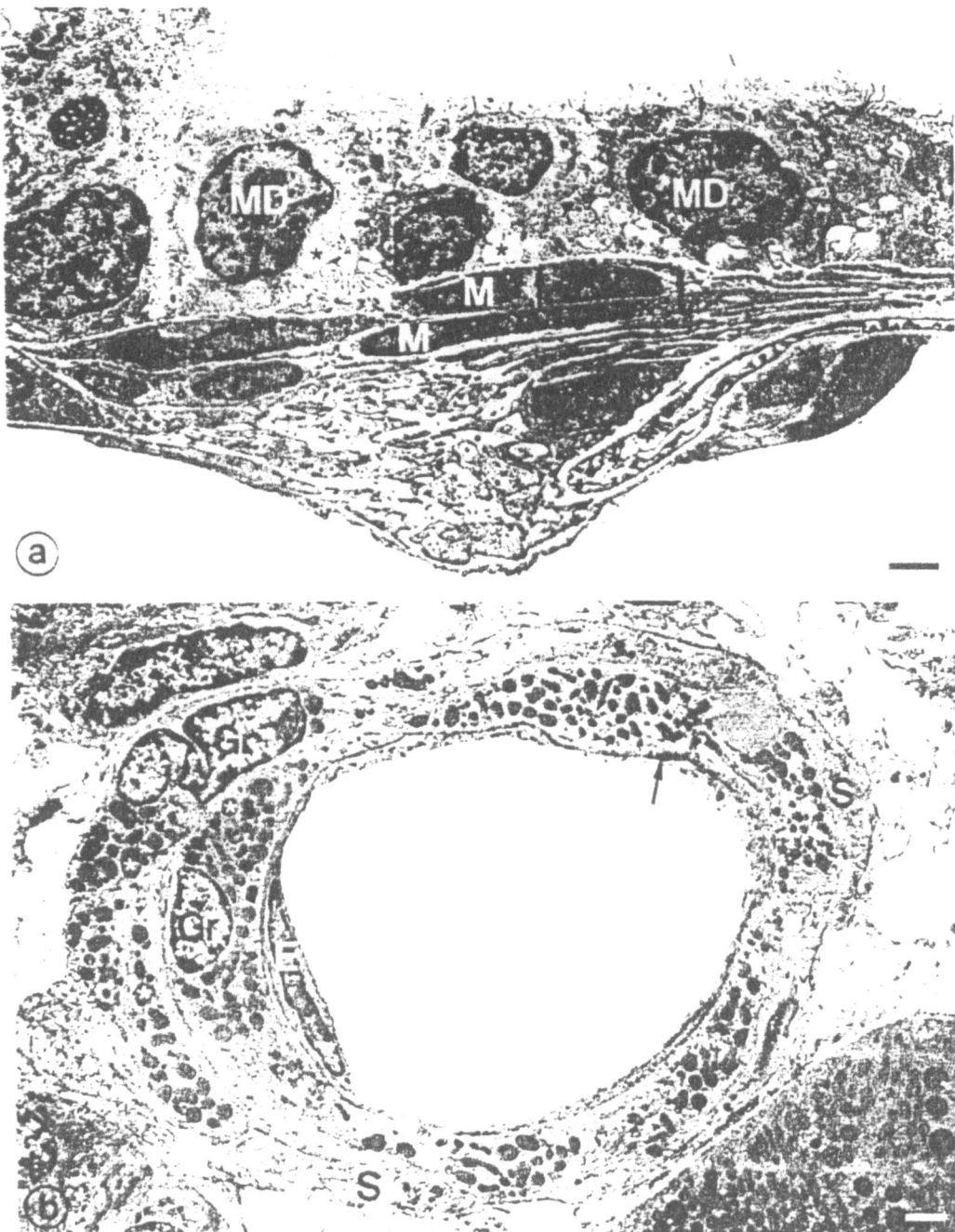

Fig. 13a, b. The juxtaglomerular apparatus. **a** Cross section through the macula densa. The lateral intercellular spaces between the macula densa cells (*MD*) are in a dilated state (*asterisks*). The base of the macula is separated by a basement membrane (*arrows*) from the extraglomerular mesangial cells (*M*) below. TEM, bar = 2 μm. **b** Cross-sectional profile through an afferent arteriole. The wall of the afferent arteriole is composed of several granular cells (*Gr*), smooth muscle cells (*S*), and an endothelial cell (*E*). Note the renin-containing granules in the granular cells (*asterisks*). Numerous gap junctions are seen as electron dense dots. TEM, bar = 2 μm

macula densa is believed to mediate the perception of the distal tubular fluid composition (NaCl concentration) and to adjust the glomerular filtration (by way of the glomerular arterioles and the mesangium) to the actual reabsorption capacity of the loop of Henle.

References

Al-Awqati Q, Schwartz GJ (1985) Plasticity in epithelial polarity. In: Frömter E (ed) 5th European colloquium on renal physiology. Frankfurt, p 7 (abstract)

Allen F, Tisher CC (1976) Morphology of the ascending thick limb of Henle. Kidney Int 9:8–22

Andrews PM, Bates SB (1984) Filamentous actin bundles in the kidney. Anat Rec 210:1–9

Ausiello DA, Kreisberg JI, Roy C, Karnovsky MJ (1980) Contraction of cultured rat glomerular cells of apparent mesangial origin after stimulation with angiotensin II and arginine vasopressin. J Clin Invest 65:754–760

Bachmann S, Kriz W (1982) Histotopography and ultrastructure of the thin limbs of the loop of Henle in the hamster. Cell Tissue Res 225:111–127

Bachmann S, Kriz W, Kaissling B (1983) Rasterelektronenmikroskopische Untersuchung kortikaler Nephronsegmente nach chronischer Furosemidapplikation. Hämodiafiltration. 16th Symposium of the Gesellschaft für Nephrologie. Salzburg, p 356 (abstract)

Bachmann S, Koeppen-Hagemann I, Kriz W (1985) Ultrastructural localization of Tamm-Horsfall glycoprotein (THP) in rat kidney as revealed by protein A-gold immunocytochemistry. In: Frömter E (ed) 5th European colloquium on renal physiology. Frankfurt, p 158 (abstract)

Baeyer H von, Deetjen P (1985) Renal glucose transport. In: Seldin DW, Giebisch G (eds) The kidney: physiology and pathophysiology. Raven, New York, pp 1663–1675

Baines AD, de Rouffignac C (1969) Functional heterogeneity of nephrons. II. Filtration rates, intraluminal flow velocities and fractional water reabsorption. Pflugers Arch 308:260–276

Barajas L (1966) The development and ultrastructure of the juxtaglomerular cell granules. J Ultrastruct Res 15:400–413

Barajas L (1970) The ultrastructure of the juxtaglomerular apparatus as disclosed by three-dimensional reconstructions from serial sections: the anatomical relationship between the tubular and vascular components. J Ultrastruct Res 33:116–147

Barajas L (1971) Renin secretion: an anatomical basis for tubular control. Science 172:484–487

Barajas L (1981) The juxtaglomerular apparatus: anatomical considerations in feedback control of glomerular filtration rate. Fed Proc 40:78–86

Barajas L, Latta H (1963) A three-dimensional study of the juxtaglomerular apparatus in the rat. Light and electron microscopic observations. Lab Invest 12:257–269

Becker B (1978) Quantitative Beschreibung der Innenzone der Rattenniere. Inaugural dissertation, Münster University

Boulpaep EL (1978) Electrophysiology of the kidney. In: Giebisch G (ed) Transport organs. Springer, Berlin Heidelberg New York, pp 97–144 (Membrane transport in biology, vol 4)

Bucher O, Kaissling B (1973) Morphologie des Juxtaglomerularen Apparates. Verh Anat Ges 67:109–136

Christensen JA, Bohle A (1978) The juxtaglomerular apparatus in the normal rat kidney. Virchows Arch [A] 379:143–150

Christensen JA, Bjoerke HA, Meyer DS, Bohle A (1979) The normal juxtaglomerular apparatus in the human kidney. A morphological study. Acta Anat (Basel) 103:374–383

Crayen ML, Thoenes W (1978) Architecture and cell structures in the distal nephron of the rat kidney. Cytobiology 17:197–211

Ernst SA (1975) Transport ATPase cytochemistry: ultrastructural localization of potassium-dependent and potassium-independent activities in rat kidney cortex. J Cell Biol 66:586–608

Ernst SA, Schreiber JH (1981) Ultrastructural localization of Na$^+$, K$^+$-ATPase in rat and rabbit kidney medulla. J Cell Biol 91:803–813

Evan A, Huser J, Bengele HH, Alexander EA (1980) The effect of alterations in dietary potassium on collecting system morphology in the rat. Lab Invest 42:668–675

Farquhar MG (1981) The glomerular basement membrane. A selective macromolecular filter. In: Hay ED (ed) Cell biology of extracellular matrix. Plenum, New York, pp 335–378

Farquhar MG, Palade GE (1962) Functional evidence for the existence of a third cell type in the renal glomerulus. Phagocytosis of filtration residues by a distinctive "third" cell. J Cell Biol 13:55–87

Farquhar MG, Wissig SL, Palade GE (1961) Glomerular permeability: I. Ferritin transfer across the normal glomerular capillary wall. J Exp Med 113:47–66

Foidart J, Sraer J, DeLarue F, Mahieu P, Ardaillou R (1980) Evidence for mesangial glomerular receptors for angiotensin II linked to mesangial cell contractility. FEBS Lett 121:333–339

Forssmann WG, Taugner R (1977) Studies on the juxtaglomerular apparatus. V. The juxtaglomerular apparatus in Tupaia with special reference to intercellular contacts. Cell Tissue Res 177:291–305

Frömter E (1979) Solute transport across epithelia: what can we learn from micropuncture studies in kidney tubules? J Physiol (Lond) 288:1–31 (The Feldberg lecture 1976)

Garg LC, Knepper MA, Burg MB (1981) Mineralocorticoid effects on Na-K-ATPase in individual nephron segments. Am Physiol 240 9:F536–F544

Goormaghtigh N (1939) Existence of an endocrine gland in the media of the renal arterioles. Proc Soc Exp Biol Med 42:688–689

Gorgas K (1978) Structure and innervation of the juxtaglomerular apparatus of the rat. Adv Anat Embryol Cell Biol 54:3–83 (English abstract)

Grantham JJ (1971) Mode of water transport in mammalian renal collecting tubules. Fed Proc 30:14–21

Grantham JJ, Burg MB (1966) Effect of vasopressin and cyclic AMP on permeability of isolated collecting tubules. Am J Physiol 211:255–259

Hansen GP, Tisher CC, Robinson RR (1980) Response of the collecting duct to disturbance of acid-base and potassium balance. Kidney Int 17:326–337

Helmchen UE (1980) Die Zahl der Mesangiumzellen in einem normalen Glomerulum der Rattenniere: eine dreidimensionale elektronenoptische Analyse. Inaugural dissertation, Tübingen University

Hoyer JR, Sisson SP, Vernier RL (1979) Tamm-Horsfall glycoprotein. Ultrastructural immunoperoxidase localization in rat kidney. Lab Invest 41:168–173

Imai M (1977) Function of the thin ascending limb of Henle of rats and hamsters perfused in vitro. Am J Physiol 232:F201–F209

Imai M, Hayashi M, Araki M, Tabei K (1984) Function of the thin limb of Henle's loop. In: Robinson RR (ed) Nephrology, vol 1. Springer, Berlin Heidelberg New York Tokyo, pp 196–207

Jamison RL, Kriz W (1982) Urinary concentrating mechanism. Structure and function. Oxford University Press, New York

Kaissling B (1980) Ultrastructural organization of the transition from the distal nephron to the collecting duct in desert rodent *Psammomys obesus*. Cell Tissue Res 212:475–495

Kaissling B (1982) Structural aspects of adaptive changes in renal electrolyte excretion. Am J Physiol 243:F211–F226

Kaissling B, Kriz W (1979) Structural analysis of the rabbit kidney. Adv Anat Embryol Cell Biol 56:1–123

Kaissling B, Kriz W (1982) Variability of intercellular spaces between macula densa cells. A transmission electron microscopic study in rabbits and rats. Kidney Int 12:S9–S17

Kaissling B, Le Hir M (1985) Anpassung distaler Tubulussegmente an Änderungen im Elektrolythaushalt. Acta Histochem [Suppl] (Jena) 31:185–192

Kaissling B, Peter S, Kriz W (1977) The transition of the thick ascending limb of Henle's loop into the distal convoluted tubule in the nephron of the rat kidney. Cell Tissue Res 182:111–118

Kaissling B, Bachmann S, Kriz W (1985) Structural adaptation of the distal convoluted tubule to prolonged furosemide treatment. Am J Physiol 248 17:F374–F381

Karnovsky MJ (1979) The structural bases for glomerular filtration. In: Jacob Churg et al. (ed) Kidney disease – present status. Williams and Wilkins, Baltimore, pp 1–41 (IAP monograph 20)

Katz AI, Doucet A, Morel F (1979) Na-K-ATPase activity along the rabbit, rat and mouse nephron. Am J Physiol 237:F114–F120

Kerjaschki D, Noronha-Blob L, Sacktor B, Farquhar MG (1984) Microdomains of distinctive glycoprotein composition in the kidney proximal tubule brush border. J Cell Biol 98:1505–1513

Kirk KL, DiBona DR, Schafer JA (1984a) Morphologic response of the rabbit cortical collecting tubule to peritubular hypotonicity: quantitative examination with differential interference contrast microscopy. J Membr Biol 79:53–64

Kirk KL, Schafer JA, DiBona DR (1984b) Quantitative analysis of the structural events associated with antidiuretic hormone-induced volume reabsorption in the rabbit cortical collecting tubule. J Membr Biol 79:65–74

Kirk KL, Bell D, Barfuss DW, Ribadeneira M (1985) Direct visualization of the isolated and perfused macula densa. Am J Physiol 248 17:F890–F894

Kriz W (1967) Der architektonische und funktionelle Aufbau der Rattenniere. Z Zellforsch Mikrosk Anat 82:495–535

Kriz W, Bankir L (1982) ADH-induced changes in the epithelium of the thick ascending limb in Brattleboro rats with hereditary hypothalamic diabetes insipidus. Ann NY Acad Sci 394:424–432

Kriz W, Schnermann J, Dieterich HJ (1972) Differences in the morphology of descending limbs of short and long loops of Henle in the rat kidney. In: Wirz H, Spinelli F (eds) Recent advances in renal physiology. Karger, Basel, pp 140–144

Kriz W, Kaissling B, Psczolla M (1978) Morphological characterization of the cells in Henle s loop and the distal tubule. In: Vogel HG, Ullrich KJ (eds) New aspects of renal function, vol 6. Excerpta Medica, Amsterdam, pp 67–78

Kriz W, Schiller A, Taugner R (1981) Freeze-fracture studies on the thin limbs of Henle's loop in *Psammomys obesus*. Am J Anat 162:23–33

Latta H, Maunsbach AB (1962) The juxtaglomerular apparatus as studied electron microscopically. J Ultrastruct Res 6:547–561

Laurie GW, Leblond CP, Inoue S, Martin GR, Chung A (1984) Fine structure of the glomerular basement membrane and immunolocalization of five basement membrane components to the lamina densa (basal lamina) and its extensions in both glomeruli and tubules of the rat kidney. Am J Anat 169:463–481

Le Hir M, Kaissling B, Dubach UC (1982) Distal tubular segments of the rabbit kidney after adaptation to altered Na⁻ and K⁻ intake. Changes in Na-K-ATPase activity. Cell Tissue Res 224:493–504

Lönnerholm G, Wistrand PJ (1984) Carbonic anhydrase in the human kidney: a histochemical and immunocytochemical study. Kidney Int 25:886–898

Madsen KM, Tisher CC (1983) Cellular response to acute respiratory acidosis in rat medullary collecting duct. Am J Physiol 245 14:F670–F679

Madsen KM, Tisher CC (1984) Response of intercalated cells of rat outer medullary collecting duct to chronic metabolic acidosis. Lab Invest 51:268–276

Marsh DJ (1970) Solute and water flows in thin limbs of Henle's loop in the hamster kidney. Am J Physiol 218:824–831

Maunsbach AB (1966) Observations on the segmentation of the proximal tubule in the rat kidney. Comparison of results from phase contrast, fluorescence and electron microscopy. J Ultrastruct Res 16:239–258

Maunsbach AB (1973) Ultrastructure of the proximal tubule. In: Orloff J, Berliner RW (eds) Handbook of physiology, vol 8. American Physiological Society, Washington, pp 31–79

Morel F, Chabardès D, Imbert-Teboul M, Le Bouffant F, Hus-Citharel A, Montégut M (1982) Multiple hormonal control of adenylate cyclase in distal segments of the rat kidney. Kidney Int [Suppl] 11:S55–S62

Neiss WF (1981) Morphogenese und Histogenese des Verbindungsstücks in der Rattenniere. Acta Anat 111:105–106 (abstract)

Peter S, Lazar J, Gross F, Forssmann WG (1974) Studies on the juxtaglomerular apparatus: II. Quantitative morphology after adrenalectomy. Cell Tissue Res 151:457–469

Pricam C, Humbert F, Perrelet A, Orci L (1974) A freeze-etch study of the tight juctions of the rat kidney tubules. Lab Invest 30:286–291

Rastegar A, Biemesderfer D, Kashgarian M, Hayslett JP (1980) Changes in membrane surfaces of collecting duct cells in potassium adaptation. Kidney Int 18:293–301

Reeves WH, Kanvar YS, Farquhar MG (1980) Assembly of the glomerular filtration surface. Differentiation of anionic sites in glomerular capillaries of newborn rat kidney. J Cell Biol 85:735–753

Rodewald R, Karnovsky MJ (1974) Porous substructure of the glomerular slit diaphragm in the rat and mouse. J Cell Biol 60:423–433

Roesinger B, Schiller A, Taugner R (1978) A freeze-fracture study of tight junctions in the pars convoluta and pars recta of the renal proximal tubule. Cell Tissue Res 186:121–133

Roll FJ, Madri JA, Albert J, Furthmayr H (1980) Codistribution of collagen types IV and AB_2 in the basement membranes and mesangium of the kidney. An immunoferritin study of ultrathin frozen sections. J Cell Biol 85:597–616

Rostgaard J, Thuneberg L (1972) Electron microscopical observations on the brush border of proximal tubule cells of mammalian kidney. Z Zellforsch 132:473–496

Rouiller C, Orci L (1971) The structure of the juxtaglomerular complex. In: Rouiller C, Muller AF (eds) The kidney: morphology, biochemistry, physiology, vol 4. Academic, New York, pp 1–80

Scherzer P, Wald H, Czaczkes JW (1985) Na-K-ATPase in isolated rabbit tubules after unilateral nephrectomy and Na^+ loading. Am J Physiol 248 17:F565–F573

Schnabel E, Kriz W (1984) Morphometric studies of the extraglomerular mesangial cell field in volume expanded and volume depleted rats. Anat Embryol (Berl) 170:217–222

Schwartz GJ, Al-Awqati Q (1985) Carbon dioxide causes exocytosis of vesicles containing H^+ pumps in isolated perfused proximal and collecting tubules. J Clin Invest 75:1638–1644

Schwartz MM, Venkatachalam MA (1974) Structural differences in thin limbs of Henle: physiological implications. Kidney Int 6:193–208

Schwartz MM, Karnovsky MJ, Venkatachalam MA (1979) Regional membrane specialization in the thin limbs of Henle's loop as seen by freeze-fracture electron microscopy. Kidney Int 16:577–589

Sikri KL, Foster CL (1981) Light and electron microscopical observations on the macula densa of the Syrian hamster kidney. J Anat 132:57–69

Silbernagel S (1985) Amino acids and oligopeptides. In: Seldin DW, Giebisch G (eds) The kidney: physiology and pathophysiology. Raven, New York, pp 1677–1701

Stanton BA, Biemesderfer D, Wade JB, Giebisch G (1981) Structural and functional study of the rat distal nephron: effects of potassium adaptation and depletion. Kidney Int 19:36–48

Stanton B, Janzen A, Klein-Robbenhaar G, DeFronzo R, Giebisch G, Wade J (1985) Ultrastructure of rat initial collecting tubule. Effects of adrenal corticosteroid treatment. J Clin Invest 75:1327–1334

Stetson DL, Wade JB, Giebisch G (1980) Morphologic alterations in the rat medullary collecting duct following potassium depletion. Kidney Int 17:45–56

Taugner R, Schiller A, Kaissling B, Kriz W (1978) Gap junctional coupling between the JGA and the glomerular tuft. Cell Tissue Res 186:279–285

Taugner R, Mannek E, Nobiling R, Bührle CP, Hackenthal E, Ganten D, Inagami T, Schröder H (1984) Coexistence of renin and angiotensin II in epithelioid cell secretory granules of rat kidney. Histochemistry 81:39–45

Trenchev P, Dorling J, Webb J, Holborrow EJ (1976) Localization of smooth muscle-like contractile proteins in kidney by immunoelectron microscopy. J Anat 121:85–95

Vasmant D, Maurice M, Feldmann G (1984) Cytoskeleton ultrastructure of podocytes and glomerular endothelial cells in man and in the rat. Anat Rec 210:17–24

Verlander JW, Madsen KM, Tisher CC (1985) Two populations of intercalated cells exist in the cortical collecting duct of the rat. Clin Res 33:501A

Wade JB, O'Neil RG, Pryor JL, Boulpaep EL (1979) Modulation of cell membrane area in renal collecting tubules by corticosteroid hormones. J Cell Biol 81P:439–445

Zalups RK, Stanton BA, Wade JB, Giebisch G (1985) Structural adaptation in initial collecting tubule following reduction in renal mass. Kidney Int 27:636–642

Cellular Mechanisms of Renal Calcium Transport

F. Lang[1], M. Paulmichl[1], and P. Deetjen[1]

Proximal Tubule

About 60% of filtered calcium is reabsorbed within the proximal convoluted tubule (Fig. 1). In the beginning of the proximal tubule (first millimeter) net calcium reabsorption lags behind net fluid reabsorption and tubular fluid calcium concentration (TF) is found to exceed that in ultrafiltrate (UF) by 10–20%. Accordingly, the ratio TF/UF approaches 1.1–1.2. In later portions of the proximal tubule calcium reabsorption parallels fluid reabsorption and TF/UF remains constant [1, 14, 17–19, 32, 46, 50, 53].

Calcium reabsorption in the proximal tubule involves both passive and active transport mechanisms.

Passive calcium transport proceeds via the paracellular shunt: Among other solutes such as chloride or sodium, calcium is dragged through the tight junctions with the bulk of reabsorbed water (solvent drag). Furthermore, an electrochemical gradient drives calcium diffusion through the paracellular shunt.

The electrical gradient depends on electrogenic transport systems and varies along the proximale tubule. In early proximal tubules, the lumen is negative and opposes calcium reabsorption. The luminal negativity comes from sodium coupled reabsorption of filtered neutral substrates such as glucose or neutral amino acids, which carries positive charge out of the lumen [43]. As a result, calcium reabsorption is relatively slow in the first portion of the proximal tubule. Furthermore, high filtered loads of substrates such as glucose stimulate sodium coupled transport in the proximal tubule and are expected to reduce proximal tubular calcium transport. As a matter of fact, infusion of glucose leads to calciuria despite the concomitant stimulation of sodium reabsorption [42].

In later portions of the proximal convoluted tubule the electrogenic sodium coupled transport gradually decays and the transepithelial potential difference shifts to lumen positive values [21]. This lumen positive potential difference is created by chloride: Along the proximal tubule the chloride reabsorption lags behind fluid reabsorption due to the preferable reabsorption of bicarbonate. As a result, luminal chloride concentration is some 20 mmol/l higher than peritubular chloride concentration and chloride following its chemical gradient, diffuses into the peritubular space leaving a lumen positive potential of some 2 mV behind [21, 22, 24].

---

[1] Institute for Physiology, University of Innsbruck, A-6010 Innsbruck, Austria.

Nephrocalcinosis, Calcium Antagonists, and Kidney
Ed. by K.-H. Bichler and W.L. Strohmaier
© Springer-Verlag Berlin Heidelberg 1988

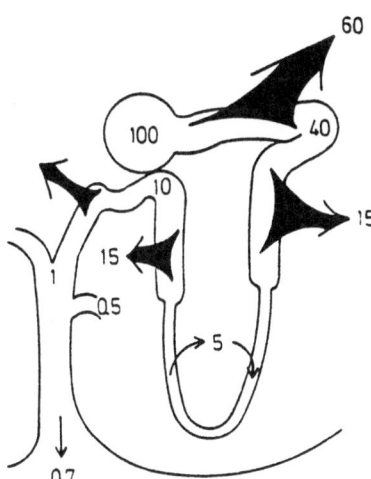

Fig. 1. Segmental reabsorption of filtered calcium (ref. see text)

The chemical gradient is created by the fluid reabsorption exceeding calcium reabsorption in the beginning of the proximal tubule resulting in a TF/UF of some 1.1 to 1.2.

Albeit small the potential difference of 2 mV and the chemical gradient of some 0.2 mmol/l drive some $^2/_3$ of proximal calcium reabsorption, since calcium permeability of the shunt is very high in the proximal tubule. The permeability coefficient as calculated from tracer calcium outflux [24] amounts to 145 $\mu m^2/s$. Thus, proximal tubular wall is almost as permeable to calcium as to sodium (155 $\mu m^2/s$) [21]. The permeability of the shunt is reduced by increasing ambient calcium concentration [23]. Passive calcium reabsorption depends on fluid reabsorption, which allows for solvent drag and builds up chemical gradients, and on bicarbonate reabsorption, which accounts for the lumen positive potential in middle to late proximal convoluted tubule. Thus proximal tubule calcium reabsorption is reduced during impaired proximal tubule fluid and/or bicarbonate reabsorption (e.g. during mannitol diuresis [17, 46] or during carbonic anhydrase inhibition [4]) and is enhanced during stimulated proximal tubular sodium reabsorption (e.g. contraction of extracellular volume).

Probably some $^1/_3$ of proximal tubular calcium transport is active.

Active calcium transport is capable to reduce luminal calcium concentration below peritubular calcium concentration [20, 46]. It proceeds via the cell [65]. Entry of calcium into the cell is driven by the steep electrical and chemical gradient at the brush border membrane. The chemical gradient comes from the difference of free calcium concentration between luminal tubular fluid (≈ 1 mmol/l) and cytosol (≈ 0.1 μmol/l [44, 47, 67]). The electrical gradient is created by the potential difference across the cell membrane, which amounts to some -70 mV (Fig. 2). The potential difference is maintained by potassium diffusion out of the cell. At the basolateral cell membrane, calcium must be transported against the same electrochemical gradient. This transport is driven by downhill movement of sodium into the cell. For extrusion of one calcium the entry of three sodium ions is needed [25, 36, 37, 44].

Calcium extrusion and transepithelial calcium transport thus depend on the electrochemical gradient for sodium. Ouabain inhibits the sodium/potassium-

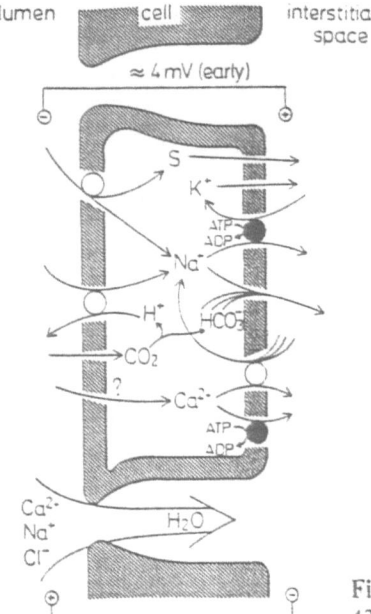

Fig. 2. Transport systems operating in the proximal tubule [6,8, 43, 44, 47]

ATPase. As a result, intracellular sodium concentration increases and the gradient for sodium dissipates. The reduction of the chemical gradient for sodium leads to impairment of calcium transport [65, 66] and intracellular concentration of free calcium increases [44, 67]. In addition to the sodium/calcium exchanger, a calcium-sensitive ATPase is found in the basolateral cell membrane [36]. Its contribution to transepithelial calcium transport is questionable, since unlike transepithelial calcium transport and intracellular calcium activity, it is not sensitive to ouabain [25, 36].

Loop of Henle

Between late proximal and early distal tubules accessible to micropuncture, some 30% of filtered calcium load is reabsorbed [13, 14]. Part of this reabsorption occurs in the pars recta [28]. Outflux of tracer calcium in that segment is similar to that in proximal convoluted tubules. Given the length of the pars recta, one might estimate that another 10–20% of filtered calcium are reabsorbed in that nephron segment.

TF/UF of calcium in thin Henle's loops of deep nephrons is 2 [35, 46] or even more [14]. Thus net fluid reabsorption obviously exceeds net calcium reabsorption in deep nephrons and/or between late proximal convoluted tubules and the bend of Henle's loops. In descending limbs the concentration of calcium increases more rapidly than that of a volume marker (inulin). Thus, some net entry of calcium probably occurs in that nephron segment [14]. As compared to the proximal tubule, the permeability of the thin limb of Henle's loop is very low. Per surface area, permeability is less than 5% of proximal permeability, per millimeter tubule length probably even less. There is no evidence for active calcium transport in that nephron segment [62].

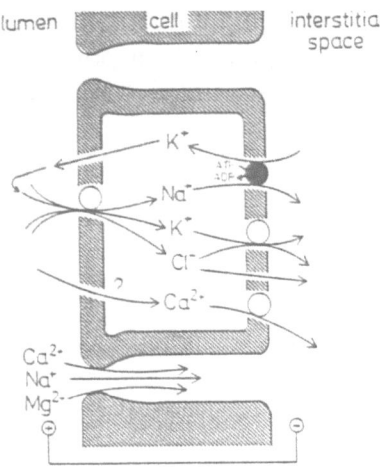

Fig. 3. Transport systems operating in the thick ascending limb [26, 29, 55]

A major portion of filtered calcium appears to be reabsorbed in the thick ascending limbs of Henle's loops [7, 34, 62, 63]. The permeability of that segment appears to be low ($\approx 10 \ \mu m^2/s$) relative to that of the proximal convoluted tubule [7, 34, 63]. However, under control conditions, efflux may exceed influx by a factor of 2–6 [7, 34, 62, 63].

In the thick ascending limb, a furosemide sensitive pump at the luminal cell membrane carries two chloride ions along with one sodium and one potassium ion into the cell (NaCl-KCl-cotransport, Fig. 3). At the basolateral cell membrane, sodium is extruded by the sodium/potassium-ATPase, in exchange for potassium. Potassium leaves the cell via a channel at the luminal cell membrane and via a coupled transport with chloride across the basolateral cell membrane. Part of chloride exit at the basolateral cell membrane is accomplished by a chloride channel [26, 29, 55]. The potassium diffusion at the luminal cell membrane and the conductive chloride exit at the basolateral cell membrane generate a lumen-positive potential of some 12 mV, which could drive calcium reabsorption. The relatively high driving force allows for substantial passive reabsorption of calcium and other cations (magnesium, sodium) despite the low permeability. Following inhibition of the NaCl/KCl cotransport by loop diuretics such as furosemide [26, 29, 55], the conductive potassium- and chloride-fluxes cease and the transepithelial potential difference vanishes. As a result, calcium reabsorption is inhibited [18, 34]. ADH, on the other hand, activates the basolateral chloride channel [26], increases the transepithelial potential difference and is expected to enhance calcium reabsorption in thick ascending limbs.

As in the proximal tubule, the permeability of the paracellular shunt is reduced by increasing ambient calcium or magnesium concentrations [15]. Thus, hypercalcemia is expected to inhibit passive sodium, calcium and magnesium reabsorption in the thick ascending limb. While other nephron segments can compensate largely for decreased calcium and sodium reabsorption, the major site for magnesium reabsorption is the thick ascending limb and hypercalcemia regularly leads to magnesuria. Conversly hypermagnesemia leads to calciuria [10, 49, 52, 61], in part due to reduction of the shunt conductance by magnesium.

Whether or not calcium reabsorption in the thick ascending limb involves an active, transcellular pathway in addition to the passive calcium transport, has been a matter of debate [7, 34, 63], which has not been resolved. Nothing is known about possible cellular transport systems.

Distal Nephron

TF/UF calcium in the distal convoluted tubule is in the range of 0.33 to 0.65 [14, 18, 46, 48]. TF/UF calcium decreases in the distal convoluted tubule [13, 46], since calcium reabsorption in this nephron segment exceeds fluid reabsorption.

Delivery of calcium out of superficial distal convoluted tubules exceeds urinary excretion [2, 13, 46, 48]. This discrepancy is suggestive of calcium reabsorption in the collecting duct. In vitro perfusion of collecting ducts showed that calcium reabsorption may occur in the granular cortical collecting duct, i.e., the connecting tubule. The light portion of the cortical collecting duct, however, is almost impermeable to calcium [63]. In addition, evidence against substantial calcium reabsorption in collecting duct comes from microinfusion studies: If tracer calcium containing fluid is microinfused into late superfical tubules, nearly 80% of injected tracer is recovered at infusion rates of 2 nl/min but close to 100% at 20 nl/min. Enhancement of infusion rate reduces contact time and thus reabsorption rate at tubular sites close to the injection site but not at sites distal to the confluence with other nephrons [27]. Since reabsorption at physiological flow rates is only 20%, reabsorption in the connecting tubule probably does not account for the entire discrepancy of calcium recovery between superficial distal tubules and urine. Instead, calcium reabsorption in deep nephrons may be more avid as in the very superficial nephrons, accessible to micropuncture. Interestingly, the calcium-reabsorbing connecting tubules of deep nephrons (arcades) appear to be extremely long as compared to connecting tubules of superficial nephrons [54].

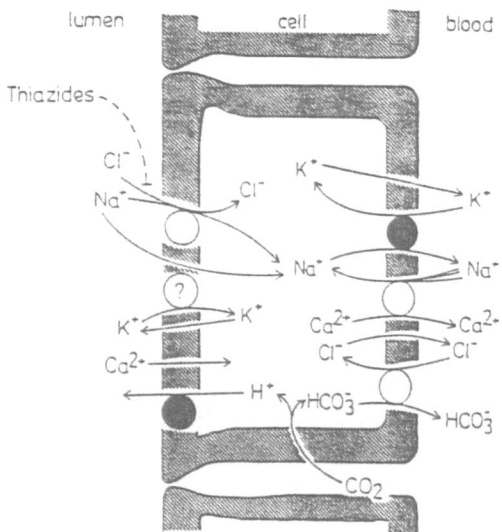

Fig. 4. Transport systems in the distal convoluted tubule [33, 39, 40]

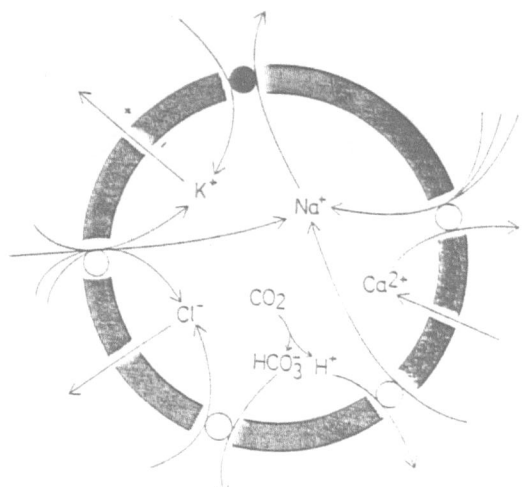

Fig. 5. Transport systems in subconfluent cultured renal epithelial cells (MDCK) [41, 45, 56, 58]

Reabsorption of calcium in distal convoluted tubules and cortical collecting duct operates both against a chemical and an electrical gradient, since electrogenic sodium reabsorption establishes a lumen-negative potential in those tubule segments (Fig. 4). Passive permeability appears to be minimal [13, 63]. Thus, all calcium reabsorption is transcellular. No direct experiments have been performed to identify cellular calcium transport systems in distal nephron segments. Some information is available from Madin-Darby-Canine-Kidney (MDCK) cells, an established cell line sharing many properties of distal tubule cells [30, 31]: In subconfluent MDCK cells, evidence is obtained for the existence of calcium channels (Fig. 5), which can seemingly be blocked by calcium antagonists [45] and stimulated by hormones such as epinephrine [57]. These calcium channels could accomplish calcium entry across the luminal cell membrane in distal tubule cells. As in MDCK cells [58] calcium could be extruded across the basolateral cell membrane of distal tubule cells in exchange for sodium.

In microperfusion experiments, no evidence for saturability of distal tubular calcium transport has been obtained [13]. However, when calcium-containing fluid is injected into free-flowing tubules, fractional calcium reabsorption decreases with increasing luminal concentration [27]. Possibly, calcium forms complexes with anions, present in distal tubular fluid but not in microperfusates, which escape reabsorption.

Calcium transport in distal convoluted tubules is closely correlated with sodium reabsorption in that segment [13], the link may be the basolateral sodium/calcium-exchanger. The correlation of sodium and calcium transport in distal convoluted tubule may be modified by several factors including parathyrin [1, 5, 9, 38, 51, 59, 64] and thiazides [3, 11, 12, 16, 18, 60]. Thiazides stimulate distal tubular calcium transport [13] despite inhibition of sodium transport. Inhibition of luminal NaCl entry by thiazides may decrease intracellular sodium activity, which should enhance sodium/calcium exchange across the peritubular cell membrane (Fig. 4). Natriuresis leads to a contraction of extracellular fluid volume, which stimulates proximal tubular fluid reabsorption. As a result, proximal calcium reabsorption is enhanced and contributes to the anticalciuria following treatment with thiazide diuretics.

References

1. Agus ZS, Gardener LB, Beck LH, Goldberg M (1973) Effects of parathyroid hormone on renal tubular reabsorption of calcium, sodium, and phosphate. Am J Physiol 224:1143–1148
2. Agus ZS, Chiu PJS, Goldberg M (1977) Regulation of urinary calcium excretion in the rat. Am J Physiol 232:F545–F549
3. Antoniou LD, Eisner GM, Slotkoff LM, Lilienfield LS (1969) Relationship between sodium and calcium transport in the kidney. J Lab Clin Med 74:410–420
4. Beck LH, Goldberg M (1973) Effects of acetazolamide and phosphate. Am J Physiol 224:1136–1142
5. Berndt TJ, Knox FG (1980) Effects of parathyroid hormone and calcitonin on electrolyte excretion in the rabbit. Kidney Int 17:473–478
6. Boron WF, Boulpaep EL (1983) Intracellular pH regulation in the renal proximal tubule of the salamander. J Gen Physiol 81:29–94
7. Bourdeau JE, Burg MB (1979) Voltage dependence of calcium transport in the thick ascending limb of Henle's loop. Am J Physiol 236:F357–F364
8. Burckhardt BC, Sato K, Frömter E (1984) Electrophysiological analysis of bicarbonate permeation across the peritubular cell membrane of rat kidney proximal tubule. Pflügers Arch 401:34–51
9. Burnatowska MA, Harris CA, Sutton RAL, Dirks JH (1977) Effects of PTH and cAMP on renal handling of calcium, magnesium, and phosphate in the hamster. Am J Physiol 233:F514–F518
10. Carney SL, Wong NLM, Quamme GA, Dirks JH (1980) Effect of magnesium deficiency on renal magnesium and calcium transport in the rat. J Clin Invest 65:180–188
11. Costanzo LS, Weiner IM (1974) On the hypocalciuric action of chlorothiazide. J Clin Invest 54:628–637
12. Costanzo LS, Weiner IM (1976) Relationship between clearances of Ca and Na: Effect of distal diuretics and PTH. Am J Physiol 230:67–73
13. Costanzo LS, Windhager EE (1978) Calcium and sodium transport by the distal convoluted tubule of the rat. Am J Physiol 235:F492–F506
14. De Rouffignac C, Morel F, Moss N, Roinel N (1973) Micropuncture study of water and electrolyte movements along the loop of Henle in psammomys with special reference to magnesium, calcium, and phosphorus. Pflügers Arch 344:309–326
15. Di Stefano A, Wittner M, Gebbler B, Greger R (1984) Increased Ca^{++} or Mg^{++} reduces the Na^+ conductance of the paracellular pathway in isolated perfused cortical thick ascending limbs of Henle loops (cTAL) of rabbit kidney. Pflügers Arch 400:R21(84)
16. Duarte CG, Bland JH (1965) Calcium, phosphorus, and uric acid clearances after intravenous administration of chlorothiazide. Metabolism 14:211–219
17. Duarte CG, Watson JF (1967) Calcium reabsorption in proximal tubule of the dog nephron. Am J Physiol 212:1355–1360
18. Edwards BR, Baer PG, Sutton RAL, Dirks JH (1973) Micropuncture study of diuretic effects on sodium and calcium reabsorption in the dog nephron. J Clin Invest 52:2418–2427
19. Edwards BR, Sutton RAL, Dirks JH (1974) Effect of calcium infusion on renal tubular reabsorption in the dog. Am J Physiol 227:13–18
20. Frick A, Rumrich G, Ullrich KJ, Lassiter WE (1965) Microperfusion study of calcium transport in the proximal tubule of the rat kidney. Pflügers Arch 286:109–117
21. Frömter E (1974) Electrophysiology and isotonic fluid absorption of proximal tubules of mammalian kidney. In: Guyton AC, Thurau K (eds) Kidney and urinary tract physiology, vol 6. Butterworths Univ Park Press, London, pp 1–39
22. Frömter E (1979) Solute transport across epithelia: What can we learn from micropuncture studies on kidney tubules? J Physiol 288:1–31
23. Frömter E, Müller CW, Knauf H (1968) Fixe negative Wandladungen im proximalen Konvolut der Rattenniere und ihre Beeinflussung durch Calciumionen. In: Watschinger B (ed) Aktuelle Probleme des Elektrolyt- und Wasserhaushalts, Nierenbiopsie. Verlag d Wiener Med Akad, 61–64

24. Frömter E, Rumrich G, Ullrich KJ (1973) Phenomenologic description of Na$^+$, Cl$^-$ and HCO$_3$$^-$ absorption from proximal tubules in the rat kidney. Pflügers Arch 343:189–220
25. Gmaj P, Murer H, Kinne R (1979) Calcium ion transport across plasma membranes isolated from rat kidney cortex. Biochem J 178:549–557
26. Greger R (1985) Ion transport mechanisms in thick ascending limb of Henle's loop of mammalian nephron. Physiol Rev 65(3):760–797
27. Greger R, Lang F, Oberleithner H (1978) Distal site of calcium reabsorption in the rat nephron. Pflügers Arch 374:153–157
28. Greger R, Lang F, Oberleithner H, Sporer H (1978) Efflux of ^{45}Calcium from proximal and distal nephron segments – effects of a diuretic (piretanide) and PTH. Pflügers Arch 377:R49
29. Greger R, Schlatter E, Lang F (1983) Evidence for electroneutral sodium chloride cotransport in the cortical thick ascending limb of Henle's loop of rabbit kidney. Pflügers Arch 396:308–314
30. Handler JS, Perkins FM, Johnson JP (1980) Studies of renal function using cell culture techniques. Am J Physiol 238:F1–F9
31. Handler JS (1983) Use of cultured epithelia to study transport and its regulation. J Exp Biol 106:55–69
32. Harris CA, Baer PG, Chirito E, Dirks JH (1974) Composition of mammalian glomerular filtrate. Am J Physiol 227:972–976
33. Hierholzer K (1985) Sodium Reabsorption in the Distal Tubular System. In: Seldin DW, Giebisch G (eds) The kidney: physiology and pathophysiology, vol 2. Raven Press, New York, pp 1063–1096
34. Imai M (1978) Calcium transport across the rabbit thick ascending limb of Henle's loop perfused in vitro. Pflügers Arch 374:255–263
35. Jamison RL, Frey NR, Lacy FB (1974) Calcium reabsorption in the thin loop of Henle. Am J Physiol 227:745–751
36. Kinne-Saffran E, Kinne R (1974) Localization of a calcium-stimulated ATPase in the basallateral plasma membranes of the proximal tubule of the rat kidney cortex. J Membr Biol 17:263–274
37. Kinne R, Keljo D, Gmaj P, Murer H (1977) The energy source of glucose and calcium transport in the renal proximal tubule. Excerpta Med Int Congr Ser No. 422, July, 1977
38. Kleeman CR, Bernstein D, Rockney R, Dowling JT, Maxwell MH (1961) Studies on the renal clearance of diffusible calcium and the role of the parathyroid glands in its regulation. Yale J Biol Med 34:1–30
39. Koeppen BM, Biagi BA, Giebisch G (1983) Intracellular microelectrode characterization of the rabbit cortical duct. Am J Physiol 244:F35–F47
40. Koeppen BM, Giebisch G, Malnic G (1985) Mechanism and regulation of renal tubular acidification. In: Seldin DW, Giebisch G (eds) The kidney: physiology and pathophysiology, vol 2. Raven Press, New York, pp 1491–1525
41. Lang F, Defregger M, Paulmichl M (1986) Apparent chloride conductance of subconfluent Madin Darby canine kidney cells. Pflügers Arch 407:158–162
42. Lang F, Joost J, Oberleithner H, Schwarz S, Pastner D (1981) Hyperglycemic calciuria. In: Calcium and phosphate transport across biomembranes. Academic Press, New York, pp 209–212
43. Lang F, Messner G, Rehwald W (1986) Electrophysiology of sodium-coupled transport in proximal renal tubules. Am J Physiol 250:F953–F962
44. Lang F, Messner G, Wang W, Oberleithner H (1983) Interaction of intracellular electrolytes and tubular transport. Klin Wochenschr 61:1029–1037
45. Lang F, Paulmichl M, Defregger M, Gstraunthaler G, Pfaller W, Deetjen P (1987) Transport systems in cultured epithelial kidney cells (MDCK) and their modulation by diuretics and hormones. In: Puschett, Greenberg (eds) Diuretics II. Elsevier Science Publishing, 107–120
46. Lassiter WE, Gottschalk CW, Mylle M (1963) Micropuncture study of renal tubular reabsorption of calcium in normal rodents. Am J Physiol 204:771–775
47. Lee CO, Taylor A, Windhager EE (1980) Cytosolic calcium ion activity in epithelial cells of Necturus kidney. Nature 287:859–861

48. Le Grimellec C, Roinel N, Morel F (1973) Simultaneous Mg, Ca, P, K, Na, and Cl analysis in rat tubular fluid. I. During perfusion of either inulin or ferrocyanide. Pflügers Arch 340:181–195

49. Le Grimellec C, Roinel N, Morel F (1973) Simultaneous Mg, Ca, P, K, Na, and Cl analysis in rat tubular fluid. II. During acute Mg plasma loading. Pflügers Arch 340:197–210

50. Le Grimellec C (1975) Micropuncture study along the proximal convoluted tubule. Electrolyte reabsorption in first convolutions. Pflügers Arch 354:133–150

51. Massry SG, Coburn JW, Chapman LW, Kleeman CR (1968) Role of serum Ca, parathyroid hormone, and NaCl infusion on renal Ca and Na clearances. Am J Physiol 214:1403–1409

52. Massry SG, Ahumada JJ, Coburn JW, Kleeman CR (1970) Effect of MgC^{12} infusion on urinary Ca and Na during reduction in their filtered loads. Am J Physiol 219:881–885

53. Morel F, Roinel N, Le Grimellec C (1969) Electron probe analysis of tubular fluid composition. Nephron 6:350–364

54. Morel F, Chabardés D, Imbert M (1976) Functional segmentation of the rabbit distal tubule by microdetermination of hormone-dependent adenylate cyclase activity. Kidney Int 9:264–277

55. Oberleithner H, Giebisch G, Lang F, Wang W (1982) Cellular mechanism of the furosemide sensitive transport system in the kidney. Klin Wochenschr 60:1173–1179

56. Paulmichl M, Gstraunthaler G, Lang F (1985) Electrical properties of Madin-Darby canine kidney cells: effects of extracellular potassium and bicarbonate. Pflügers Arch 405:102–107

57. Paulmichl M, Defregger M, Lang F (1986) Effects of epinephrine on electrical properties of Madin-Darby canine kidney cells. Pflügers Arch 406:367–371

58. Paulmichl M, Friedrich F, Lang F (1986) Electrical properties of Madin Darby canine kidney cells: effects of extracellular sodium and calcium. Pflügers Arch 407:258–263

59. Peacock M, Robertson WG, Nordin BEC (1969) Relation between serum and urinary calcium with particular reference to parathyroid activity. Lancet I:384–386

60. Quamme GA, Wong NLM, Sutton RAL, Dirks JH (1975) Interrelationship of chlorothiazide and parathyroid hormone: a micropuncture study. Am J Physiol 229:200–205

61. Quamme GA, Dirks JH (1980) Intraluminal and contraluminal magnesium on magnesium and calcium transfer in the rat nephron. Am J Physiol 238:F187–F198

62. Rocha AS, Magaldi JB, Kokko JP (1977) Calcium and phosphate transport in isolated segments of rabbit Henle's loop. J Clin Invest 59:975–983

63. Shareghi GR, Stoner LC (1978) Calcium transport across segments of the rabbit distal nephron in vitro. Am J Physiol 235:F367–F375

64. Sutton RAL, Wong NLM, Dirks JH (1976) Effects of parathyroid hormone on sodium and calcium transport in the dog nephron. Clin Sci Molecul Med 51:345–351

65. Ullrich KJ, Rumrich G, Klöss S (1977) Phosphate transport in the proximal convolution of the rat kidney. I. Tubular heterogeneity, effect of parathyroid hormone in acute and chronic parathyroidectomized animals and effect of phosphate diet. Pflügers Arch 372:269–274

66. Vogel G, Stockert I (1967) Die Bedeutung des Natriums für den renaltubulären Calcium-Transport bei Rana ridibunda. Pflügers Arch 298:23–30

67. Wang W, Messner G, Oberleithner H, Lang F, Deetjen P (1984) The effect of ouabain on intracellular activities of K^-, Na^+, Cl^-, H^+ and Ca_2^- in proximal tubules of frog kidney. Pflügers Arch 401:6–13

68. Windhager EE, Taylor A (1983) Regulatory role of intracellular calcium ions in epithelial Na transport. Ann Rev Physiol 45:519–532

Factors Influencing Renal Calcium Excretion

J. H. DIRKS[1]

Nephrocalcinosis is associated with a number of clinical disorders such as primary hyperparathyroidism, renal tubular acidosis, sarcoidosis, vitamin D toxication and the milk alkalii syndrome. In all of these disorders there are potential derangements of factors that normally regulate renal tubular calcium reabsorption. Thus, it is important to gain understanding of the possible reasons for nephrocalcinosis and these other disorders by reviewing the basic pathophysiology of calcium transport by the renal tubule. The basic mechanisms of renal calcium transport in various nephron segments have already been reviewed by Deetjeen. This review will assess the various physiological and pharmacological factors that can affect renal calcium transport at various nephron segments [1, 2].

Hemodynamic Factors and Changes in Extracellular Fluid Volume

As glomerular filtration rate (GFR) is acutely increased or reduced there is a roughly proportional increase or decrease in calcium excretion as well as that for sodium. As chronic renal failure ensues with an irreversible reduction in GFR fractional calcium excretion may initially be reduced reflecting enhanced tubular reabsorption likely due to secondary hyperparathyroidism. As renal failure becomes more advanced the absolute calcium excretion remains low due to the very low filtered load but fractional calcium excretion, like that of sodium, becomes higher and the tendency to normal calcium excretion is maintained. This reflects an impairment of distal calcium reabsorption relative to that of GFR despite high PTH levels which occur in advanced chronic renal failure. In our own experiments in the remnant kidney model of azotemia, proximal tubule and loop fractional calcium reabsorption became depressed accounting for the higher fractional calcium excretion observed.

Extracellular fluid volume has a major affect on calcium excretion as it does for sodium chloride and other ions. With saline infusion or other forms of volume expansion the clearance of filterable calcium increases in parallel with that of sodium in a progressive manner. This occurs even in the presence of reduced filtered load reflecting that the depression of tubular reabsorption which is due to progressively marked decreases in sodium calcium reabsorption in the proximal tubule with a resetting downward of distal and likely collecting duct system reab-

---

[1] Department of Medicine, University of British Columbia, Vancouver, BC, U5Z 1M9, Canada.

Nephrocalcinosis, Calcium Antagonists,
and Kidney
Ed. by K.-H. Bichler and W. L. Strohmaier
© Springer-Verlag Berlin Heidelberg 1988

sorption of calcium. Our own studies indicated that as the flow-rate through the thick ascending limb of Henle's loop increased, this of itself reduced the positive potential difference across the epithelium due to diminishing transepithelial sodium gradient and would reduce the passive reabsorption of calcium across the tight junctions. This would deliver more calcium to the distal tubule and saturate distal tubule and collecting duct calcium transport yielding greater calcium excretion. The converse situation of extracellular volume contraction modestly reduces calcium excretion through overall enhancement of salt and water reabsorption in the proximal tubule delivering less filtered calcium to the distal nephron and leading to a decrease in calcium excretion.

Ionic Effects

Ionic effects play a major role in renal tubular calcium reabsorption when one considers changes in serum calcium, phosphate, magnesium, and a variety of anions.

Hypercalcemia and Hypocalcemia

Hypercalcemia is a major factor in the genesis of nephrocalcinosis. The effects of hypercalcemia on glomerular filtration rate and tubular reabsorption are complex. First of all, as serum calcium rises a reduction in GFR is often noted. Acute infusions of calcium salts can reduce both single nephron and whole kidney GFR with a fall in the glomerular capillary ultrafiltration coefficient. This appears to occur only if PTH is present which also causes a decrease in GFR and glomerular ultrafiltration coefficient. The dominant effects of hypercalcemia relate to a decrease in tubular calcium reabsorption at several levels. In the proximal tubule there is decrease in bulk salt and water, along with calcium reabsorption with an additional selective impairment of calcium reabsorption in extreme hypercalcemia. In the thick ascending limb of Henle's loop calcium reabsorption is progressively decreased in hypercalcemia with a simultaneous fall in magnesium reabsorption. This occurs particularly when calcium was raised on the peritubular side and not on the luminal side of the membrane. The mechanisms of these effects are still unclear but may relate to competition for a common calcium-magnesium transport site on the basolateral cellular membrane or this may relate to changes in paracellular permeability reducing passive calcium as well as magnesium reabsorption. Distal tubule and collecting duct reabsorption of calcium appears also to be decreased. This occurs in part due to suppression of PTH due to hypercalcemia but it also appears likely that there is additional inhibiting effect of hypercalcemia per se on "distal" tubular calcium reabsorption. This still requires further experimentation.

   Hypocalcemia results in decreased calcium excretion first because of the lower filtered load and the proportional reduction ion tubular reabsorption throughout the nephron. There is also evidence that hypocalcemia per se also augments calcium reabsorption especially in the thick ascending limb. The release of PTH due to hypocalcemia in the presence of functioning parathyroid glands would, of course, increase calcium transport in the thick ascending limb and the distal tubule. However, there would appear to be specific changes due to hypocalcemia itself in increasing calcium transport in the loop of Henle.

Phosphate alters calcium reabsorption. In micropuncture studies, hyperphosphatemia indicate that some enhanced calcium reabsorption occurs in the distal tubule and the collecting duct system. By contrast phosphate depletion is a cause of hypercalciuria resulting from impaired tubular calcium reabsorption. This would appear to be not fully accounted by PTH suppression and an additional role of phosphate depletion in reducing calcium reabsorption may occur due to a reduction of proximal tubule fluid reabsorption as well as a more distal effect. We had previously observed that in phosphate depleted dogs the distal tubule calcium concentration was higher than normal suggesting a defect prior to the puncture site probably in the thick ascending limb. Correction of phosphate depletion with phosphate infusions restored calcium reabsorption at a site beyond the distal tubule. The suggestion has been made that the defect in calcium transport due to phosphate depletion occurred in the distal tubule and collecting duct system.

A variety of anions when infused intravenously can cause calcium complexes with a disproportionate increases in urinary calcium excretion. Anions such as sulphate, ferracyanide and gluconate can do this. Microperfusion studies in the rat with sulphate reduce calcium reabsorption in the proximal tubule, loop and distal segments with probably the greatest effect in the distal tubule.

Magnesium can markedly alter calcium excretion, hypermagnesemia due to infusion of magnesium salts increases the urinary excretion of magnesium, sodium, and calcium. In our own studies of hypermagnesemia using micropuncture techniques indicated that overall proximal tubule salt and water, as well as magnesium and calcium reabsorption was reduced to a modest degree. A more dramatic decrease in tubular transport occurred in the thick ascending limb in that a profound decrease in magnesium reabsorption occurred with concurrent reduction in calcium reabsorption. Thus, a very significant calcium load is delivered to the distal tubule which is only partially reabsorbed back so that hypercalciuria resulted. These studies again speak of an interaction between calcium and magnesium transport in the thick ascending limb so that high concentrations of either ion from the peritubular side can reduce each other's reabsorption with a relatively greater effect on magnesium transport. The converse situation of hypomagnesemia have not been as well worked out but the evidence suggests that both magnesium and calcium reabsorption are relatively enhanced in the thick ascending limb. This may have relevance in the disorder of familial hypercalcemic hypocalciuria in which there is both hypercalcemia and hypermagnesemia and in which the relationship between tubular calcium and magnesium may be disrupted resulting in greater calcium and magnesium reabsorption at any filtered load. It should also be mentioned that chronic lithium administration has a distal calcium retaining effect.

Acid Base Alterations

There is now considerable evidence that metabolic acidosis causes hypercalciuria and metabolic alkalosis causes hypocalciuria. In particular, chronic metabolic acidosis has been shown to be associated with decreased tubular calcium reabsorption as hypercalciuria occurs despite a diminished filtered load. In our own studies using micropuncture techniques in the dog made acidotic by feeding ammonium chloride, calcium reabsorption is impaired relative to sodium at the distal tubular micropuncture site whereas reabsorption remained in parallel in the proximal convoluted tubule. Correction of the acidosis with acute infusions of sodium bicarbonate in both parathyroid intact and thyroparathyroidectomized acidotic dogs restored this impairment of calcium reabsorption and in fact enhanced calcium reabsorption over and above control. These data suggest that a distal calcium transport site presumably including the distal tubule and the thick ascending limb in the calcium reabsorptive system which is inhibited by chronic metabolic acidosis and increased by metabolic alkalosis perhaps directly related to increased fluid bicarbonate delivery. In more recent studies using in vitro microperfusion of the straight proximal tubule of the rabbit we have also demonstrated an effect of acid base changes on calcium transport. In these studies acidosis reduced net calcium transport and this occurred by an increase of calcium backflux from the interstitium into the lumen. Conversely, imposition of alkalosis increased net calcium transport by reducing the backflux into the lumen [3]. Thus, changes in metabolic alkalosis may be effective at several nephron sites including the straight portion of the proximal tubule and likely the thick ascending limb and the distal tubule. Effects of acidosis and alkalosis are independent of the action of parathyroid hormone as the effects can be equally well shown in thyroparathyroidectomized animals. Effects of acidosis and alkalosis are important in understanding conditions of nephrocalcinosis. It would appear that in the milk alkali syndrome, a very strong factor in the promotion of enhanced calcium renal tubular reabsorption would be alkalosis itself. Changes in tubular hydrogen ion content may also be important in the nephrocalcinosis associated with renal tubular acidosis.

Hormonal Effects

Calcium reabsorption is affected by a number of hormones but particularly by parathyroid hormone as well as by the vitamin D metabolites. Other hormones have lesser effects which are less important in the normal physiological control of renal calcium reabsorption.

Parathyroid Hormone

The calcium retaining effect of parathyroid hormone is well known and considerable research have been performed in the last number of years to delineate the mechanism of action. Parathyroid hormone has multiple renal effects. First of all, as the circulating concentration rises, it may decrease GFR by primarily reducing glomerular capillary ultrafiltration coefficients. The dominant effects of PTH though are in enhancing tubular calcium reabsorption and we were able to show this quite decisively in the golden hamsters in which the effect of PTH upon renal calcium handling is particularly large. In the proximal tubule, PTH has been observed to have varied responses with decreases, increases, or no change in calcium reabsorption. It would appear that the PTH effect on proximal calcium reabsorption, unlike its effects on phosphate reabsorption, is not responsible for the final urinary hypocalciuria. In the thick ascending limb of Henle's loop calcium reabsorption has been shown to be enhanced and this effect is mediated by cyclic AMP. The nature of the effect in thick ascending limb may be dependent on an alteration in transepithelial positive electrical voltage. It has been suggested that the effects of PTH may result from alteration of the permeability of calcium channels in the tight junctions in the thick ascending limb. Other studies have also suggested the direct effect of PTH on active calcium transport in the thick ascending limb. Further studies in both free-flow micropuncture and isolated microperfusion of the distal tubule have indicated enhanced calcium reabsorption due to PTH at this site with an additional effect on the terminal nephron segments. Again, the effects were mediated by cyclic AMP. Thus, as a generalization the tendency of PTH is to decrease proximal calcium reabsorption and enhance calcium reabsorption in the ascending limb, distal tubule, and perhaps the terminal nephron segments yielding in the final urine as relative hypocalciuria for the particular filtered load.

Vitamin D Metabolites

The effects of vitamin D metabolites upon renal calcium transport are complex. In the vitamin D depleted dog and rat, vitamin D metabolite appears to enhance tubular calcium reabsorption. Recently. Yamamoto and his colleagues [4] observed in parathyroparathyroidectomized rats with vitamin D deficiency is associated with decreased tubular calcium reabsorption and the PTH responsiveness of calcium reabsorption was reduced. This suggested that vitamin D and PTH may act cooperatively to enhance tubular calcium reabsorption and help to raise the serum calcium level. In the vitamin D replete parathyroidectomized dogs infused with saline and ADH all of the vitamin D metabolites caused acute enhancement of tubular calcium reabsorption as well as sodium and phosphate. In micropuncture studies, the effect of 25-dihydroxy D3 in increasing calcium reabsorption appear to be located in the distal nephron. By contrast, in the vitamin D repleted rat physiological doses of 1,25-dihydroxy D3 did not affect renal tubular calcium reabsorption. In the vitamin D replete hamster in which baseline calcium excretion is very sensitive to PTH 1,25-dihydroxy D3 appeared to antagonize the effects of low-dose PTH infusion and result in hypercalciuria. In the absence of PTH or full replacement of PTH the antagonistic effects of 1,25-hydroxy D3 were not apparent. This complex set of observations makes it difficult to reach a final conclusion but there is the suggestive evidence of opposing effects of PTH and vitamin D in some species. It is to be noted that the distal convoluted and connecting tubule contains a vitamin D dependent calcium binding protein where active calcium transport occurs but this calcium binding protein is absent from the proximal tubule, the loop of Henle where active calcium transport does not appear to occur. The mechanism of nephrocalcinosis in vitamin D toxication would presumably be dependent on the increase in the filtered load of calcium due to the hypercalcemia from increased intestinal reabsorption bringing about an increased absolute reabsorption of calcium by the nephron. This may be furthered by direct effects of the vitamin D metabolites in enhancing tubular calcium reabsorption above normal.

To briefly consider other hormones chronic glucocorticoid excess as occurs in Cushing's syndrome results in hypercalcemia presumably due to the overall renal response to the resorption from the skeleton increasing filtered load. The acute administration of mineralcorticoids does not alter urinary calcium excretion but as mineralcorticoid escape from the sodium retention occurs calcium excretion progressively increases above the original control values. This can be prevented by sodium restriction during the period of sodium retention which is caused by mineralcorticoids. It has been suggested that mineralcorticoids can also enhance calcium secretion ion into the cortical collecting tubules. Calcitonin would also appear to be hypocalciuric and this effect has been observed to occur in the thick ascending limb and distal tubule. Glucose administration can be accompanied by hypercalciuria due to the reduced tubular reabsorption. Insulin infusion has a similar response. Micropuncture studies have shown that hyperglycemia with insulin infusions cause inhibition of proximal calcium reabsorption independent of PTH. Growth hormone excess such as occurs in acromegaly causes hypercalciuria. This appears to be a tubular effect but the exact site has not been identified. Urinary calcium secretion is increased in hyperthyroidism and decreased in hypothyroidism. It is uncertain whether the thyroid hormones affect tubular calcium handling directly. Estrogens reduce calcium excretion but this may be secondary to decrease at bone reabsorption. Prolactin excess can cause hypercalciuria and recent micropuncture evidence suggest that this occurs in the distal tubule where prolactin depresses calcium transport. Other hormones such as glucagon and vasopressin may increase calcium excretion due to their effects on reducing calcium transport the thick ascending limb. The overall, quantitative importance of such a variety of hormones on renal calcium transport would appear much less than factors such as parathyroid hormone, alkalosis, and serum calcium concentration.

Diuretics

Much has been written about the effects of various diuretics on calcium transport and these diuretics have been used extensively therapeutically. Carbonic anhydrase inhibitors such as acetazolamide inhibit proximal tubular reabsorption of sodium bicarbonate profoundly and also reduce calcium reabsorption but because of compensation in the more "distal" nephron there is little change in calcium excretion. Furosemide and other loop diuretics produce major increases in calcium excretion as they do for other cations. These effects are greatest during the acute diuretic period and when triggered with volume expansion. The major site of action is in the thick ascending limb of Henle's loop and the reabsorption of sodium, calcium, and magnesium declines proportionately. Normally, the absorption of these ions is in large part secondary to lumen positive transepithelial potential difference generated by the sodium chloride reabsorption resulting from the luminal sodium potassium chloride cotransport system. Furosemide blocks this luminal transport system, reduces the positive transepithelial voltage, and reduces the passive backflux of calcium across the tight junctions. Thus, sodium, calcium, magnesium, and potassium reabsorption are inhibited in parallel when loop diuretics are applied. Calcium excretion may be enhanced more than that of sodium and this may reflect greater reabsorption of sodium back into the circulation than occurs for calcium in the distal tubule. The acute calciuretic affect of furosemide is used in the treatment of symptomatic hypercalcemia. Extracellular fluid volume contraction is carefully prevented with saline infusions and high fractional excretion of urinary calcium can be maintained with frequent doses of furosemide. Care is taken to avoid general depletion of other ions, especially potassium and magnesium.

The thiazide diuretics have the interesting opposite effect of enhancing calcium reabsorption in the distal convoluted tubule and this has become a mainstay of the treatment of recurrent renal calculii with hypercalciuria. The acute effects of thiazide diuretics in animal studies leads to an increase in sodium excretion with little exchange in calcium excretion. Our micropuncture studies of the dog show that this acute effect occurred between the proximal and distal tubule sampling sites and furthermore, that this effect was independent of the presence of PTH. Our studies in the hamster confirmed the distal tubule as the site of these calcium containing effects due to thiazide. Further, Costanza and Windhagger showed in distal microperfusion studies that the acute hypercalciuric effect of chlorothiazide decisively occurred in the distal convoluted tubule where sodium reabsorption was decreased while calcium reabsorption was being enhanced. There was no change in the distal transepithelial potential difference in response to the thiazide diuretics. The mechanism of enhanced distal calcium reabsorption due to thiazides is unknown but it is independent of parathyroid hormone. The chronic hypocalciuric effect of thiazides have been very important in the treatment of many patients with hypercalciuria and recurrent renal stones.

The natriuretic potassium sparing diuretics such as amiloride are also hypocalciuric. Amiloride decreases distal tubule and collecting duct electrical potential nd enhances calcium reabsorption. In the isolated profused collecting tubule, evidence has been presented to show that amiloride decreased passive calcium secretion into the lumen. As the effects of amiloride and thiazide are on different nephron segments, the calcium retention may be additive and can be used to advantage. Osmotic diuretics such as mannitol can cause profound increase in calcium excretion as well as that of other cations. These effects occur due to reduction in overall ion reabsorption of the proximal tubule with greater reduction in ion reabsorption in the loop of Henle due to the rapid flow rate through the thick ascending limb reducing the favourable passive gradients for reabsorption.

Summary

The factors that affect renal tubular calcium reabsorption are summarized in Table 1. Under normal circumstances, less than 2% of the filtered calcium is excreted in the urine. The renal tubular reabsorption of calcium is involved in retrieving most of the filtered load with the capacity to fine tune calcium excretion according to physiological needs. Reabsorption of the bulk of filtered calcium closely parallels that of sodium and this occurs in the proximal convoluted and the loop of Henle so that only about 10% of the filtered load of these ions reaches the distal convoluted tubule. The adjustment of urinary calcium losses occurs principally in the final nephron segments, the distal convoluted, and collecting duct system. Here calcium reabsorption is promoted by parathyroid hormone, metabolic alkalosis, probably vitamin D metabolites and high phosphate levels. In the terminal nephron segments or collecting ducts metabolic acidosis and probably phosphate depletion inhibit calcium reabsorption and cause hypercalciuria. Key therapeutic agents influence calcium reabsorption with loop diuretics reducing it in the thick ascending limb and thiazide diuretics and amiloride increasing calcium reabsorption in the distal convoluted tubule and cortical collecting ducts respectively. The ability to enhance calcium reabsorption by parathyroid hormone, alkalosis, and vitamin D metabolites obviously play a crucial role in the generation of nephrocalcinosis.

Table 1. Summary of factors affecting fractional renal tubular calcium reabsorption

	Increase	Decrease
Hemodynamic volume related	Increased filtered load	Decreased filtered load – acute and chronic
	Volume concentration	Volume expansion
Ionic	Hypocalcemia Hypomagnesemia Hyperphosphatemia Lithium	Hypercalcemia Hypermagnesemia Phosphate depletion
Acid base	Metabolic alkalosis	Metabolic acidosis
Hormonal	Parathyroid hormone Vitamin D metabolites	Hypoparathyroid ? Vitamin D metabolite when D repletion and submaximal amounts of PTH *Misc:* prolactin, growth hormone, glucagon. vasopressin
Diuretics	Thiazides Amiloride	Loop

References

1. Sutton RAL, Dirks JH (1986) Calcium and magnesium: renal handling and disorders of metabolism. In: Brenner BM, Rector FCR (eds) The kidney. Saunders, Philadelphia, pp 551–618
2. Bourdeau J (1983) Renal handling of calcium. In: Brenner BM, Stein JH (eds) Contemporary issues in nephrology, divalent ion homeostasis. Churchill Livingstone, New York Edinburgh London Melbourne
3. Wong NLM, Dirks JH (1985) Differential effects of acid-base changes on proximal straight tubule (PSt) transport of calcium and magnesium. Submitted to Federation Proceedings 44:1914
4. Yamamoto M, Kawanobe Y, Ogata E (1983) In vivo renal Ca transport in vitamin D-deficient rats. Abstracts of the VIIIth International Conference on Calcium Regulating Hormones, Kobe, Japan. P123

Ca²⁺ Antagonists – Mode of Action and Pharmacodynamics

H. D. LEHMANN[1]

In 1969 Albrecht Fleckenstein coined the term Ca^{2+} antagonist in order to define the action of verapamil, gallopamil and prenylamine [12]. Today the term Ca^{2+} antagonist (or Ca^{2+} entry blocker or slow channel blocker) is being used for a whole group of substances which basically have the same mode of action despite their heterogenous chemical structure [31]. The group of Ca^{2+} antagonists comprises

- phenylalkylamine derivatives with their prototype verapamil,
- dihydropyridine derivatives with their prototype nifedipine,
- benzothiazepine derivatives with their prototype diltiazem (Fig. 1, Table 1)

These substances have been very thoroughly investigated. In addition there are also substances (e.g.. flunarizine, bepridil and fendiline) which have only a few features in common with the above mentioned Ca^{2+} antagonists and are thus not discussed here.

Mode of Action

The common feature of Ca^{2+} antagonists is that they block Ca^{2+} influx through specific channels into the cell. These Ca^{2+} channels are protein pores situated in the phospholipid bilayer of the plasma membrane (Fig. 2). They can be opened by phosphorylation via a cAMP-dependent protein kinase, activated during depolarization of the cell.

Through one channel three million Ca^{2+} ions can enter the cardiac muscle cell per second. This number is fairly small, compared to the influx of Na^+ through Na^+ channels in the same cell (which is 50 times higher), and that is why the Ca^{2+} channels are also named slow channels. Though not exclusive, their selectivity for Ca^{2+} is at least one hundred times greater than for Na^+ or K^+. However, under experimental conditions where there is either no Ca^{2+} or the concentration is very low the Na^+ flux becomes predominant [34, 44].

These Ca^{2+} channels open when the membrane potential is reduced during depolarization which means that the process is voltage dependent (voltage-operated channels). By measuring the Ca^{2+} flux in individual Ca^{2+} channels using the patch-clamp technique [36, 16] it was shown that these channels do not open at

[1] Knoll AG, Biological Research and Development, Department of Cardiovascular Pharmacology, Knollstr., D-6700 Ludwigshafen, FRG.

Nephrocalcinosis, Calcium Antagonists, and Kidney
Ed. by K.-H. Bichler and W.L. Strohmaier
© Springer-Verlag Berlin Heidelberg 1988

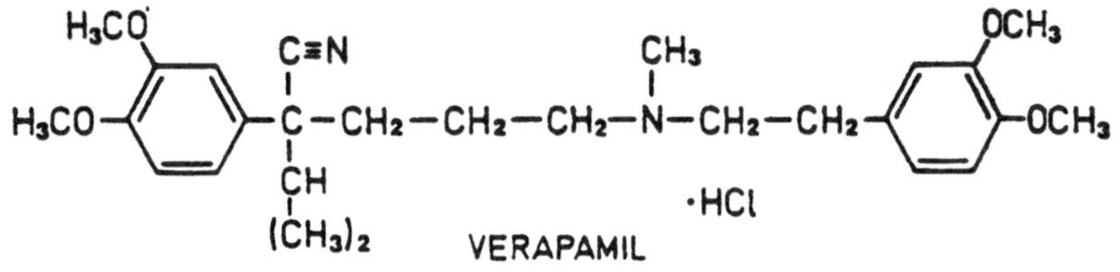

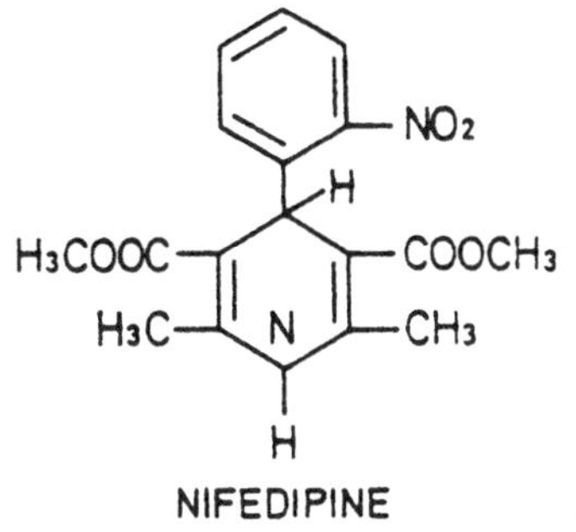

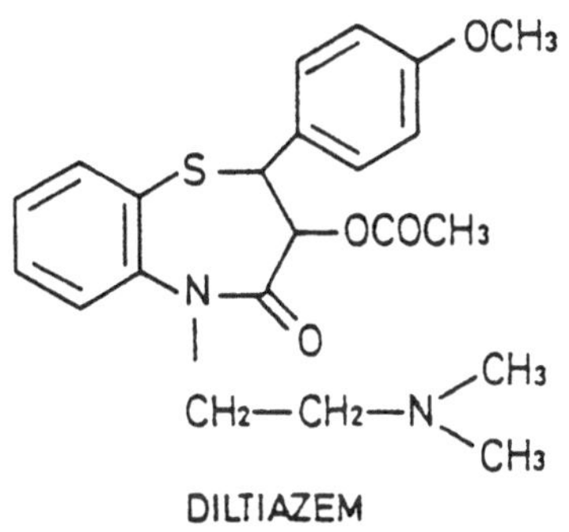

Fig. 1. Chemical structures of verapamil, nifedipine, and diltiazem

Table 1. Survey of Ca^{2+} antagonists

Compounds	Source
Phenylalkylamines	
Verapamil	Knoll AG (Germany)
Gallopamil	Knoll AG (Germany)
Anipamil	Knoll AG (Germany)
Devapamil	Knoll AG (Germany)
Ronipamil	Knoll AG (Germany)
Emopamil	Knoll AG (Germany)
Tiapamil	Hoffmann-La Roche (Switzerland)
Dihydropyridine derivatives	
Nifedipine	Bayer AG (Germany)
Niludipine	Bayer AG (Germany)
Nitrendipine	Bayer AG (Germany)
Nimodipine	Bayer AG (Germany)
Nicardipine	Yamanouchi (Japan)
Felodipine	Haessle/Astra (Sweden)
Darodipine	Sandoz (Switzerland)
Benzothiazepine derivatives	
Diltiazem	Tanabe (Japan)

Extracellular space
$[Ca^{2+}]\ 10^{-3}M$

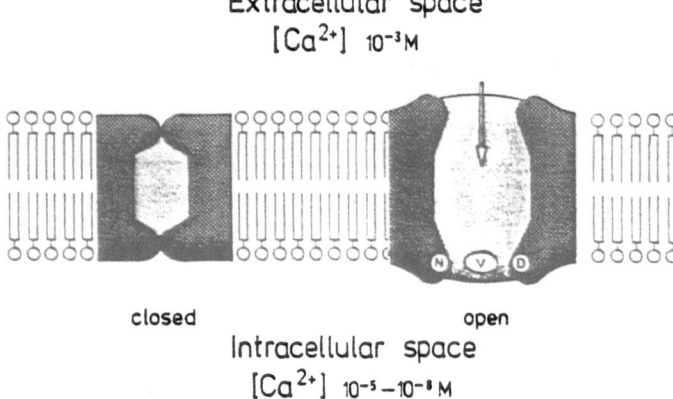

closed   open

Intracellular space
$[Ca^{2+}]\ 10^{-5}-10^{-8}M$

Fig. 2. Scheme of the voltage-operated Ca^{2+} channels in the cell membrane with three different binding sites for the Ca^{2+} antagonists verapamil (V), nifedipine (N) and diltiazem (D)

fixed intervals but follow a statistical distribution pattern. The channels open for about 1 millisecond and then close again for periods from one to several hundred milliseconds. Physiologically, neurotransmitters, and catecholamines in particular, modulate this process by prolonging the time the channels stay open and shortening the time they are closed. It has not yet been fully explained whether this modulation also effects the number of Ca^{2+} channels [43]. Ca^{2+} antagonists have the opposite effect: they shorten the open time and prolong the closed time of Ca^{2+} channels and thus reduce the Ca^{2+} influx; that is why they are also called slow channel or Ca^{2+} entry blockers [42].

Voltage-operated channels are found everywhere in the organism, e.g., in myocardial cells, in smooth muscle cells, in the impulse generation and conduction system of the heart, and in skeletal muscle cells. In fact, the t-tubule system in the skeletal muscle cell has so many of these channels that it is the experimental organ most often used for binding studies [15]. Recently some investigators succeeded in isolating Ca^{2-} channels from the tissue and embedding them in an artificial membrane [14]. Thus, it is now possible to investigate how these channels function and the parameters which influence them.

In order to be able to isolate these Ca^{2+} channels from the membranes, the channel constituents had to be marked. For this purpose Ca^{2+} antagonists which bind specifically to these channels were used. The different chemical structures of the phenylalkylamine derivative, the dihydropyridine derivatives, and the benzothiazepine derivative suggests that the various Ca^{2+} antagonists bind to different receptor sites. By using highly specific tritiated ligands $((-)-[^3H]$ devapamil for the phenylalkylamine receptor, $(+)-[^3H]$ PN 200-110 for the dihydropyridine receptor) it has been shown that the three groups of Ca^{2+} antagonists really have three different drug receptor sites which are close to one another within the membrane. If one of the receptor sites is already taken this will have a positive or negative effect on the binding of the other ligands [15, 41, 35]. These receptors are believed to be localized at the cytoplasmic side of the channel. By marking different tissues with tritiated Ca^{2+} antagonists, specific receptor sites have been found not only in the heart, blood vessels, and skeletal muscles but also in the brain, adrenal cortex, red cells, and other tissues (Table 2). In particular the electric eel (*Electrophorus electricus*), has many such binding sites. These are presumed to be

Table 2. Distribution of saturable [^{3}H]nimodipine binding sites [15]

Species	Tissue	Specifically bound ligand (fmol/mg protein)
Guinea-pig	Kidney	31
	Medulla/Pons	42
	Liver	48
	Thalamus/Midbrain	68
	Hypothalamus	82
	Striatum	90
	Lung	106
	Cerebellum	119
	Duodenum	124
	Uterus	152
	Cerebral cortex	153
	Olfactory tubercle	163
	Hippocampus	200
	Adrenal	212
	Heart	371
	Skeletal muscle	1770
Hamster	Fat cell membranes	0
Human	Platelet membranes	0
Human	Erythrocyte ghosts	26
Electrophorus electricus	Membranes from the electric organ	1680

linked to the ability of this South American fresh water fish to generate high voltages of up to 800 V [15].

We now know that Ca^{2+} antagonists bind specifically to cell membranes, to structures that are Ca^{2+} channels, but we still have much to learn about Ca^{2+} channels themselves and their biological functions. For instance, we do not know exactly what the function of Ca^{2+} channels in the skeletal muscle tubules, the preferred organ for binding studies, actually is. Nor do we know to what extent skeletal muscle cells are affected by Ca^{2+} antagonist binding. However, we are already aware of the fact that in these cells the voltage-operated Ca^{2+} channels constitute a heterogeneous group in which some can be blocked by Ca^{2+} antagonists and others can not [7].

Furthermore there is at least a second type of Ca^{2+} channel in the plasma membrane in which the Ca^{2+} flux is primarily triggered by various transmitters (receptor-operated Ca^{2+} channels) [4, 28] and where no change in the membrane potential is required. The opening process of receptor-operated Ca^{2+} channels (Fig. 2) is most likely controlled by the formation of inositol triphosphate, IP_3 (from phosphoinositol diphosphate, PIP_2) which functions as an intracellular messenger [2, 3]. There is an interaction between receptor-operated Ca^{2+} channels and voltage-operated Ca^{2+} channels but this has been poorly understood to date [40]. In general, Ca^{2+} antagonists seem to act less on receptor-operated Ca^{2+} channels than on voltage-operated Ca^{2+} channels [6]. In this context it may be of importance to mention that verapamil, gallopamil, and diltiazem also bind to α-adrenoceptors and, in this way, can change membrane function [1, 11, 30].

Pharmacodynamics

Of all parenchymal cells, the myocardial cell is the one which has been studied most extensively with respect to transmembrane Ca^{2+} flux and how it is affected by Ca^{2+} antagonists. The myocardial cell membrane has about 30 channels per square micrometer. Between 5 and 10 µmol Ca^{2+}/kg heart weight flow into the cells per beat [21]. This accounts for only about one-tenth of the Ca^{2+} required to activate contraction. However, the incoming Ca^{2+} ("trigger Ca^{2+}") determines the quantity of Ca^{2+} released from intracellular stores (Fig. 3). In myocardial cells the Ca^{2+} stores in the sarcoplasmic reticulum are more important than the Ca^{2+} stores in the mitochondria. Thus, the force of myocardial contraction is closely linked to the extent of transmembrane Ca^{2+} influx through the Ca^{2+} channels during depolarization. In cells other than myocardial cells (e.g., platelets, neutrophils, hepatocytes, adrenal glomerulose cells) the increase in cytoplasmic Ca^{2+} concentration, which is necessary for cell activation, is caused through receptor-coupled IP_3 formation resulting in Ca^{2+} influx from the extracellular space and Ca^{2+} release from the endoplasmic reticulum [2, 3]. In order to maintain intracellular Ca^{2+} homeostasis, pumps which extrude or sequestrate Ca^{2+} from the cytoplasm are required. These Ca^{2+} pumps are located in the plasma membrane, the sarcoplasmic reticulum, and the endoplasmic reticulum (Fig. 3).

The influx of Ca^{2+} into myocardial cells is reflected in the action potential, the shape and duration of its plateau phase depending on the Ca^{2+} influx (Fig. 4). Thus the action potential is a very sensitive parameter which can be recorded without great technical problems. Ca^{2+} antagonists dose-dependently shorten the plateau phase of the action potential and decrease the contractile force (Fig. 5). The unchanged upstroke velocity of the action potential, which is related to Na^+ influx through fast channels, indicates that this effect is selective. The fact that an increase in extracellular Ca^{2+} concentration suppresses a Ca^{2+}-antago-

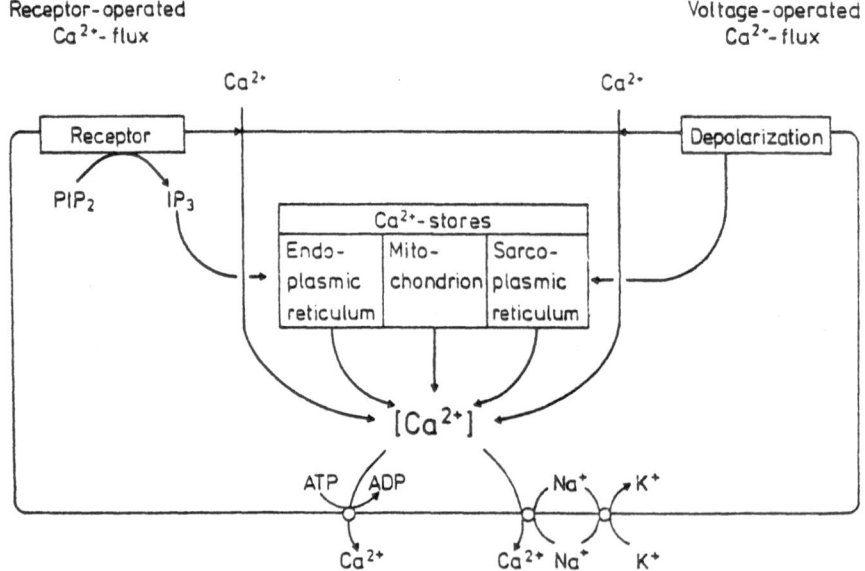

Fig. 3. Mechanisms for regulating intracellular calcium

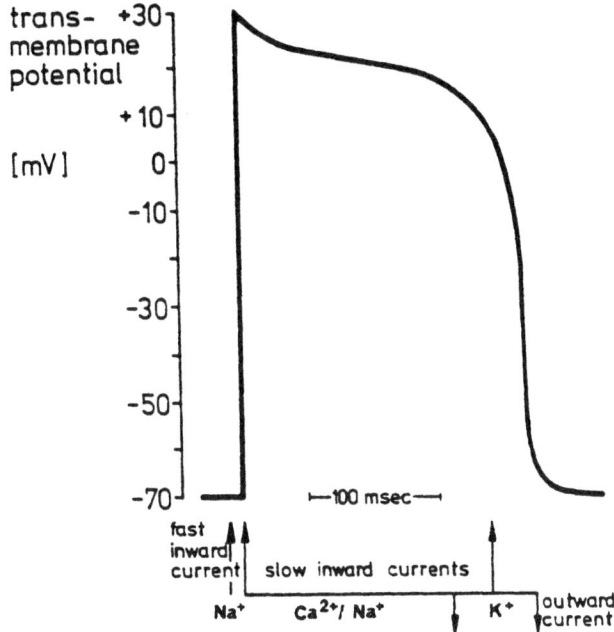

Fig. 4. Ionic fluxes during ventricular action potential

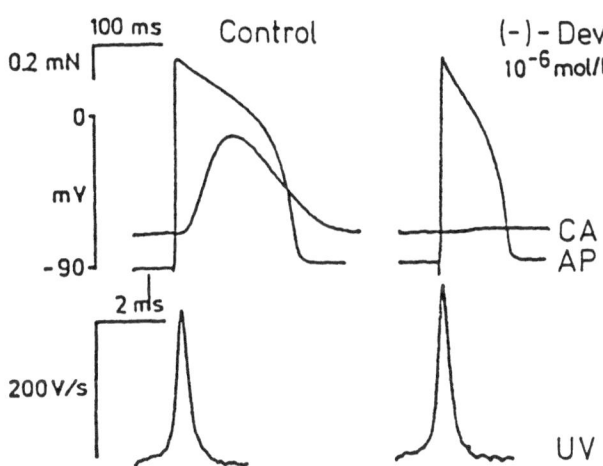

Fig. 5. Influence of (−)-devapamil on the action potential (*AP*) and contraction amplitude (*CA*) of guinea-pig papillary muscle. The upstroke velocity (*UV*) remains unchanged [29]

nistic effect is further experimental evidence for a Ca^{2+}-antagonistic mode of action [19]. This is noncompetitive antagonism since Ca^{2+} and Ca^{2+} antagonists do not compete for the same receptor.

The Ca^{2+} antagonistic effect on myocardial cells is used therapeutically to protect the cells in conditions where there is a pathologically increased Ca^{2+} influx (Ca^{2+} overload) into the cell. These conditions may be due, for example, to a massive release of catecholamines or occur during the reperfusion phase following cell-damaging ischemia. The negative chronotropic and dromotropic activities of Ca^{2+} antagonists like verapamil are also of therapeutic importance. These effects result from the fact that electrical activity of SA-node and AV-node cells largely depends on slow channel transport.

The effect of Ca^{2+} antagonists on parenchymal cells other than those of the heart (brain, liver, and kidney cells) were also investigated. This paper only deals

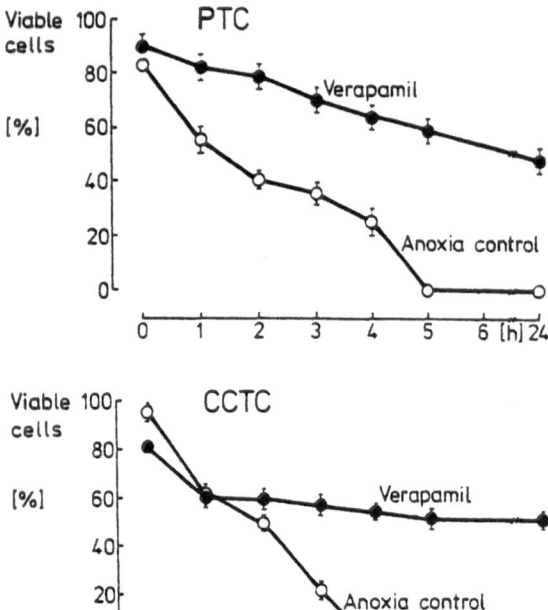

Fig. 6. Beneficial effect of verapamil on in vitro renal cell ischemia [38]. (*PTC*, Proximal tubules cells; *CCTC*, cortical collecting tubule cells)

with the results obtained in renal cells. Schwertschlag et al. report experiments conducted on cultures of proximal tubule cells and collecting tubule cells under anoxic conditions and on the influence of extracellular Ca^{2+} decrease and the use of Ca^{2+} antagonists [38]. If these cells are rendered anoxic for 45 min at 25 °C they will be damaged and, even after transferring them to a normoxic medium, all the cells die within 4–6 h. The cells' survival is assessed by their ability to eliminate nigrosine dye. It is possible to slow down the dying off of these cells either by reducing the extracellular Ca^{2+} concentration in the postanoxic phase or by adding Ca^{2+} antagonists to the medium. Therefore Ca^{2+} plays a major role in renal cell damage during the postanoxic reflow phase. At verapamil concentrations of 5×10^{-8} mol/l the cell survival rate at 5 h after anoxia is 60% while all the control cells have died by that time (Fig. 6). At this concentration, which also corresponds to the effective Ca^{2+}-antagonistic concentrations in the myocardial cell, verapamil prolongs the survival time of renal cells from 5 to more than 24 h. This direct effect of Ca^{2+} antagonists on renal tubule cells must be considered part of their well-known action on the kidney, in addition to their influence on renal vascular tone in hypoxic and anoxic experimental conditions and in human kidney transplantations [8, 9, 18, 20, 39].

There is a great number of experiments reporting on the Ca^{2+} antagonists' relaxing effect on vascular smooth muscle cells (e.g., [23]). The Ca^{2+}-antagonistic activity can best be shown in vitro on vessel strips (e.g., of the rat aorta), first depolarized by high extracellular K^+ concentrations, then contracted by adding Ca^{2+}. This is an experimental method of inducing voltage-operated Ca^{2+} influx [19]. The mean concentrations ($EC_{50\%}$) of some Ca^{2+} antagonists required to inhibit rat aorta contractions triggered as described above are summarized in Table 3. Clinically, the vasodilator effect of Ca^{2+} antagonists is used to dilate cor-

Table 3. Inhibitory activity ($IC_{50\%}$) of Ca^{2+} antagonists on K^+-induced contraction of rat aorta [33]

Ca^{2+} antagonist	$IC_{50\%}$ (mol/l)
Phenylalkylamine	
Verapamil	3.5×10^{-8}
Gallopamil	1.4×10^{-8}
Devapamil	4.0×10^{-9}
Dihydropyridine	
Nifedipine	2.6×10^{-9}
Nimodipine	6.7×10^{-10}
Nicardipine	6.5×10^{-10}
Benzothiazepine	
Diltiazem	2.3×10^{-7}

onary vessels and reduce afterload in coronary heart disease, to treat spasms in cardiac and cerebral vessels, and to lower blood pressure in hypertension.

Ca^{2+} antagonists have proved effective antihypertensive drugs both in animal models and in man [5]. It is very likely that in this case Ca^{2+} antagonists have a causal mechanism of action and that hypertension is the result of an increased transmembrane Ca^{2+} turnover. Experiments on rats have shown that the potency of Ca^{2+} antagonists depends on the existing blood pressure level and also on the type of hypertension involved (Fig. 7). In order to lower the blood pressure with verapamil in the normotensive rat, relatively large doses are needed, the $ED_{20\%}$ being 39 mg/kg orally [32]. Of the three types of experimentally induced hypertension, namely that in the spontaneously hypertensive rat, the stroke-prone spontaneously hypertensive rat, and the desoxycorticosterone acetate salt (DOCA salt) hypertensive rat, the latter type responds most readily with an $ED_{20\%}$ of 13 mg/kg orally. These results indicate that the Ca^{2+} influx may be considerably increased under pathological conditions.

In animal experiments the antihypertensive effect of Ca^{2+} antagonists is accompanied by a reduction of the calcium content of the vascular wall and a prolongation of the animal's life span [13, 17]. Furthermore recent experiments have provided some evidence that, in the vascular system, Ca^{2+} antagonists not only act on vascular smooth muscle cells but also on endothelial cells, thus inhibiting macromolecular leakage [25, 26].

What is known about the action of Ca^{2+} antagonists on blood constituents? Studies in coronary patients have shown that during treatment with Ca^{2+} antagonists pathologically reduced red cell deformability was significantly improved [45]. Recently, by adding gallopamil to patients' blood in vitro, Ernst and Matrai [10] were able to restore pathologically impaired filtering capacity and reduce red cell aggregation. These experiments demonstrated the direct effect of the Ca^{2+} antagonist on these cells. It seems that Ca^{2+} influx in red cells is increased under certain pathological conditions.

Rats (Strain)	Initial values (mmHg)	ED 20°/₀ [mg/kg] 2 h
Normotensive ● (Sprague-Dawley)	148 ± 2.3	39
Spontaneously hypertensive ○ (Okamoto)	237 ± 4.5	25
Stroke prone spont. hyperten. ■ (Okamoto-Yamori)	255 ± 4.8	17
DOCA / Salt-hypertensive □ (Sprague-Dawley)	248 ± 4.6	13

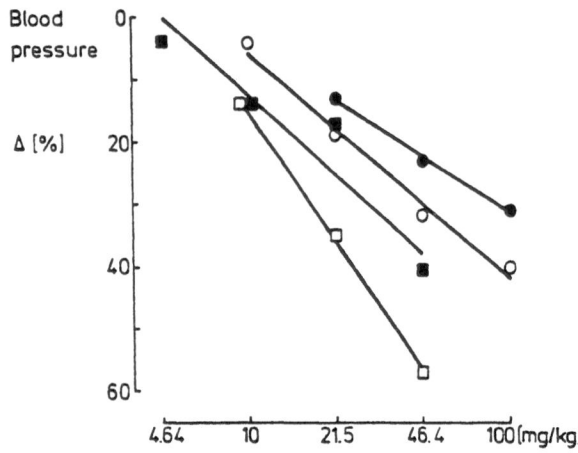

Fig. 7. Comparison of the blood pressure lowering effect (ED$_{20\%}$) in normotensive and hypertensive rats after oral administration of verapamil [32]

Many papers deal with the effect of Ca^{2+} antagonists on platelet aggregation [27]. According to Glossmann these cells have no specific receptor sites for Ca^{2+} antagonists [15]. At fairly low concentrations, verapamil and gallopamil have been found to have a very good inhibitory effect on noradrenaline and platelet-aggregating factor (PAF) induced aggregation [22, 24]. We may conclude that this effect is closely related to their binding to α_2-adrenoceptors and PAF receptors.

It is remarkable that the Ca^{2+} antagonist doses required to inhibit platelet aggregation in vivo are much smaller than those required in vitro. In our studies, we determined a mean inhibitory concentration (EC$_{50\%}$) of 15 mg/l for verapamil on combined adrenaline-collagen aggregation of canine platelets in vitro, while as little as 1 mg/kg was needed to produce the same effect when giving the substance orally [22]. This phenomenon which has also been found for nimodipine on other platelet parameters is still unclear [24]. It is possible that the Ca^{2+} turnover in vivo is more marked and thus can be manipulated more easily than in vitro. It is also conceivable that the platelets on passing through the splanchnic vessels (in which there are much higher Ca^{2+} antagonist concentrations after enteral absorption than occur in the systemic circulation after liver passage) undergo a lasting change in Ca^{2+} turnover. This would explain why there is no correlation between systemic plasma levels and the inhibitory effect.

Conclusion

In summary it is obvious that Ca^{2+} antagonists are effective in different parts of an organ in vivo, namely in the parenchymal cells, in the cells of the vessel wall, and finally in the blood constituents. The fact that the Ca^{2+} antagonists have such a large number of possible sites of action in the organism opens up the possibility of a great variety of therapeutic uses, many of which have yet to be discovered. Remarkable new effects and fields of application should be found in the future.

References

1. Barnathan ES, Addonizio VP. Shattil SJ (1982) Interaction of verapamil with human platelet α-adrenergic receptors. Am J Physiol 242:H19–H23
2. Berridge MJ (1984) Inositol triphosphate and diacylglycerol as second messengers. Biochem J 220:345–360
3. Berridge MJ, Irvine RF (1984) Inositol triphosphate, a novel second messenger in cellular signal transduction. Nature (London) 312:315–321
4. Bolton TB (1979) Mechanisms of action of transmitters and other substances on smooth muscle. Phys. Rev 59:606–718
5. Bühler FR (1985) Kalziumantagonisten. In: Ganten D, Ritz E (Hrsg) Lehrbuch der Hypertonie. Schattauer. Stuttgart. S 686–694
6. Cauvin C. Loutzenhiser R, Van Breemen C (1983) Mechanisms of calcium antagonist-induced vasodilation. Ann Rev Pharmacol Toxicol 23:373–396
7. Cognard Ch, Lazdunski M, Romey G (1986) Different types of Ca^{2+} channels in mammalian skeletal muscle cells in culture. Proc Natl Acad Sci USA 83:517–521
8. Duggan KA, MacDonald GJ. Charlesworth JA, Pussell BA (1985) Verapamil prevents post-transplant oliguric renal failure. Clin Nephrol 24:289–291
9. Eisinger DR, Suranyi MG, Bracs P, Farnsworth A, Sheil AGR (1985) Effects of verapamil in the prevention of warm ischemia induced acute renal failure in dogs. Aust NZJ Surg 55:391–396
10. Ernst E, Matrai A (1986) Beeinflussung der Erythrozytenrheologie durch Gallopamil in vitro. In: Tillmann W, Ehrly AM (eds) Hämorheologie und Hämatologie. MWP Verlag, München, pp 146–148
11. Fairhurst AS, Whittaker ML, Ehlert FJ (1979) Interactions of D600 (methoxy verapamil) and local anaesthetics with rat brain α-adrenergic and muscarinic receptors. Biochem Pharmacol 29:155–162
12. Fleckenstein A. Tritthart H, Fleckenstein B, Herbst A, Grün G (1969) Eine neue Gruppe kompetitiver Ca^{++}-Antagonisten (Iproveratril, D 600, Prenylamin) mit starken Hemmeffekten auf die elektromechanische Koppelung im Warmblüter-Myokard. Pflügers Arch 307: Suppl R 25
13. Fleckenstein A (1983) Calcium antagonism in heart and smooth muscle. Wiley, New York
14. Flockerzi V, Pelzer D, Cavalié A, Trautwein W. Hofmann F (1986) Solubilization and reconstitution of calcium channels. Naunyn-Schmiedeberg's Arch Pharmacol 332: Suppl R 43
15. Glossmann H, Ferry DR, Goll A, Striessnig J, Zernig G (1985) Calcium channels and calcium channel drugs: recent biochemical and biophysical findings. Arzneim-Forsch/Drug Res 35(II):1917–1935
16. Franciolini F (1986) Patch clamp technique and biophysical study of membrane channels. Experientia 42:589–594
17. Gries J, Feilner K (1984) Untersuchungen zur blutdrucksenkenden Wirkung von Anipamil, einem neuen Ca-Antagonisten. Therapiewoche 34:6390

18. Hertle L, Garthoff B (1985) Calcium channel blocker nisoldipine limits ischemic damage in rat kidney. J Urol 134:1251–1254
19. Hof RP, Vuorela HJ (1983) Assessing calcium antagonism on vascular smooth muscle: a comparison of three methods. J Pharmacol 9:41–52
20. Kramer HJ, Neumark A, Schmidt S, Klingmüller D, Glänzer K (1983) Renal functional and metabolic studies on the role of preventive measures in experimental acute ischemic renal failure. Clin Exp Dial Apheresis 7:77–99
21. Langer GA (1980) The role of calcium in the control of myocardial contractility: an update. J Mol Cell Cardiol 12:231–239
22. Lehmann HD (1984) New evidence of a potent inhibitory activity on platelet aggregation by verapamil. Naunyn-Schmiedeberg's Arch Pharmacol 325: Suppl R 45
23. Lipe S, Moulds JD (1983) Calcium, drug action and hypertension. Pharmacol Ther 22:299–330
24. MacIntyre DE, Shaw AM (1983) Phospholipid-induced human platelet activation: effects of calcium channel blockers and calcium chelators. Thromb Res 31:833–844
25. Mayhan WG, Joyner WL (1984) The effect of altering the external calcium concentration and a calcium channel blocker, verapamil, on microvascular leaky sites and dextran clearance in the hamster cheek pouch. Microvasc Res 28:159–179
26. McDonagh PF, Roberts DJ (1986) Prevention of transcoronary macromolecular leakage after ischemia-reperfusion by the calcium entry blocker nisoldipine. Circ Res 58:127–136
27. Mehta JL (1985) Influence of calcium-channel blockers on platelet function and arachidonic acid metabolism. Am J Cardiol 55:158B–164B
28. Meisheri KD, Hwang O, Van Breemen C (1981) Evidence for two separate calcium pathways in smooth muscle plasmalemma. J Membr Biol 59:19–25
29. Nawrath H, Raschack M (1984) Calcium antagonistic effects of the radioligand (–)-desmethoxy-verapamil on cardiac and vascular smooth muscle preparations. Cell Calcium 5:316
30. Nayler WG, Thompson JE, Jarrott B (1982) The interaction of calcium antagonists (slow channel blockers) with myocardial alpha adrenoceptors. J Mol Cell Cardiol 14:185–188
31. Nayler WG, Horowitz JD (1983) Calcium antagonists: a new class of drugs. Pharmacol Ther 20:203–262
32. Raschack M, Gries J, Kirchengast M, Bühler V (1987) Chronic oral verapamil prevents progression of hypertension and prolongs life-span in spontaneously hypertensive rats. In: Fleckenstein A, Laragh (eds) Hypertension – the next decade: verapamil in focus. Churchill Livingstone, Edinburgh London Melbourne New York, pp 302–306
33. Raschack M (unpublished results)
34. Reuter H (1979) Properties of two inward membrane currents in the heart. Ann Rev Physiol 41:413–424
35. Ruth P, Flockerzi V, Oeken H-J, Hofmann F (1986) Solubilization of the bovine cardiac sarcolemmal binding sites for calcium channel blockers. Eur J Biochem 155:613–620
36. Sakmann B, Neher E (eds) (1983) Single-channel recording. Plenum Press, New York
37. Schrör K, Latta G, Darius H, Klaus W, Ziegler R (1985) Hemmung der Plättchenaggregation und Thromboxanbildung durch den Calcium-Antagonisten Nisoldipin nach einer oralen Einmaldosis von 10 mg. Klin Wochenschr 63:16–19
38. Schwertschlag U, Schrier RW, Wilson P (1986) Beneficial effects of calcium channel blockers and calmodulin binding drugs on in vitro renal cell anoxia. J Pharmacol Exper Therap 238:119–124
39. Shapiro J, Cheung C, Itabashi A, Chan L, Schrier R (1985) The protective effect of verapamil on renal function after cold ischemia in the isolated perfused rat kidney. Transplantation (Baltimore) 40:596–600
40. Siegel G, Adler A (1985) The effect of noradrenaline on membrane potential and tension in vascular smooth muscle. Pflügers Arch 403: Suppl R 58
41. Striessnig J, Goll A, Moosburger K, Glossmann H (1986) Purified calcium channels have three allosterically coupled drug receptors. FEBS 197:204–210
42. Trautwein W, Cavalié A (1985) Cardiac calcium channels and their control by neurotransmitters and drugs. J Am Coll Cardiol 6:1409–1416

43. Trautwein W, Pelzer D (1985) Voltage-dependent gating of single calcium channels in the cardiac cell membrane and its modulation by drugs. In: Marmé D (ed) Calcium physiology. Springer, Berlin Heidelberg New York Tokyo, pp 53–93
44. Tsien RW (1983) Calcium channels in excitable cell membranes. Ann Rev Physiol 45:341–358
45. Waller DG, Nicholson HP, Roath S (1984) The acute effects of nifedipine on red cell deformability in angina pectoris. Br J Clin Pharmacol 17:133–138

Nephrocalcinosis and Calcium Antagonists

Pathological Anatomy, Etiology, and Pathogenesis of Nephrocalcinosis

H. G. LABERKE[1]

Introduction

Since the first description of nephrocalcinosis (NC) by Virchow in 1855, not much attention has been paid to this disease from a morphological point of view; only a few fundamental light and electron microscopic studies are available (Randall and Melvin 1937; Jaccottet 1957; Caulfield and Schrag 1964; Anderson 1982), but the physiology and pathophysiology of the renal handling of calcium have evoked increasing interest.

In the following the wide range of NC is presented with respect to the possible protective and therapeutic (?) effects of calcium antagonists. I intend to give exact definitions, to document light and electron microscopic findings critically with regard to the literature, and to give detailed information on etiology and pathogenesis. In addition, the reversibility of NC and the relationship between NC and nephrolithiasis will be discussed, although the pathologist cannot relate the application of calcium antagonists to the different types and primary diseases of NC.

Definitions

Nephrocalcinosis

Primary = metastatic NC: due to general dysfunction of the metabolism; calcification of nondamaged tissue (Jaccottet 1959; Scarpelli 1965; Reubi 1970; Cottier 1980).
Secondary = dystrophic NC: calcification following necrosis of renal parenchyma or of deposited abnormal material in the kidney (Reubi 1970; Cottier 1980).

The differentiation between primary and secondary NC is by no means always possible; however, this distinction should be maintained for didactic reasons. We assume that primary NC can develop rather quickly into secondary NC if functional or structural parenchyma damage promotes further calcifications.

In accordance with Anderson (1982) and Heptinstall (1983) we call *any* deposition of calcium in renal tissue, NC.

---

[1] Institute of Pathology, University of Tübingen, Liebermeisterstr. 8, D-7400 Tübingen, FRG.

Nephrocalcinosis, Calcium Antagonists, and Kidney
Ed. by K.-H. Bichler and W.L. Strohmaier
© Springer-Verlag Berlin Heidelberg 1988

Hypercalciuria (Hautmann and Lutzeyer 1980)

Absorptive hypercalciuria: due to excessive intestinal calcium absorption.
Renal hypercalciuria: increased calcium excretion as a result of impaired fine regulation of calcium handling in the distal nephron.
Resorptive hypercalciuria: due to excessive bone resorption, for example in primary hyperparathyroidism.
Idiopathic hypercalciuria: unknown cause (about 20% of all cases).

Nephrolithiasis (Cottier 1980)

Macroscopically visible pelvicalyceal concrements.
Primary nephrolithiasis: metabolic causes, idiopathic cases.
Secondary nephrolithiasis: result of special local conditions like urinary tract infections, congenital or acquired urinary tract obstruction.

Table 1. Morphologic methods for the identification of (cell) calcium

---

A. *Light microscopic methods*
  1. *Substitution methods*
     - Silver stain method of Kossa
  2. *Calcium detection in mineralised tissue*
     - Hematoxylin-eosin staining
     - Murexide
     - Calcon
     - Calcein
     - Anthraquinon derivates
       - Alizarine
       - Quinalizarine
       - Alizarin red S
       - Nuclear fast red (calcium red)
     - Sodium rhodizonate
     - Phthalocyanine
     - Ferrocyanide
     - Naphthalhydroxamic acid
  3. *Vital staining of calcium in tissues*
     - Tetracycline
     - Alizarin red S
     - Fluorexon
  4. *Detection of ionized or ionizable calcium*
     - Morine (fluorescence microscopy)
     - Alizarin red S (histochemistry)
     - Glyoxal-bis-(2-hydroxyanil) [GBHA]

B. *Electron microscopic detection of calcium*
  - Conventional fixed tissue
  - Osazone method
  - Oxalate method
  - Pyroantimonate method

C. *Radiologic microanalysis*

D. *Autoradiography*

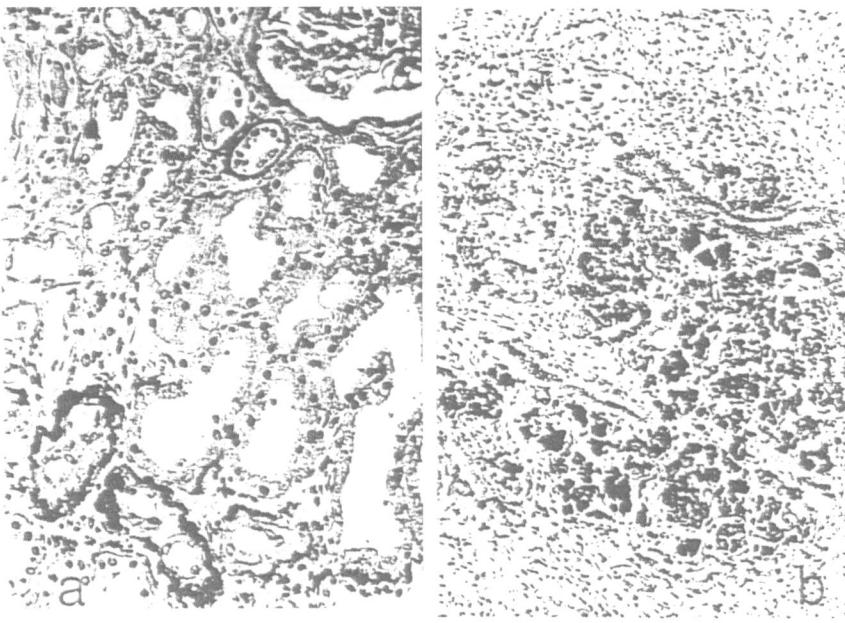

Fig. 1 a, b. Demonstration of calcium deposits (Kossa reaction). **a** Black-stained densely positioned granular calcium (salts) in the basement membrane of atrophic renal tubules. × 160. **b** Plaque-shaped calcification of damaged tubules in the medullary region, × 80

Methods of Calcium Detection

Table 1 presents the different methods of calcium detection. The routinely applicated Kossa reaction (Kossa 1901) does not identify calcium itself but its salt component; this substitution method is very reliable (Bunting 1951; Schäfer 1979). The black staining allows identification of even very small quantities of calcium salts (Fig. 1). In contrast the conventional PAS reaction in routine kidney histology is not suitable for detecting smaller calcifications. This, however, is possible using hematoxylin-eosin staining.

Electron microscopic findings should be considered, as they might be important for the evaluation of the pathogenesis of NC. The other methods listed in Table 1 will not be discussed in detail.

As deposits of iron can also be found in the area of the calcium salt deposits a follow-up stain of all my cases using the Berlin-blue reaction was done; this yielded total negative results.

Incidence of Nephrocalcinosis

Light microscopically, NC is found in 0.9% of the cases in biopsy material (Zollinger and Mihatsch 1978); in autopsy material the values range from about 20% to 100% (Ungar 1950; Epstein 1971; Haggitt and Pitcock 1971; Weller et al. 1972;

Anderson 1982). Magori et al. (1983), however, report a very high incidence in biopsy material following an electron microscopic study: they showed in 78 of 196 biopsy cases the presence of calcium phosphate at the proximal tubular basement membrane and in Bowman's capsule. They interpreted this as early dystrophic calcification.

When applying a comprehensive definition of NC the above-mentioned figures reported in autopsy material should be regarded as too low. Assuming an accurate examination and the most appropriate staining method, calcification may be found anywhere in almost every kidney.

Etiology of Nephrocalcinosis

In Table 2 I have listed the different etiological factors of NC, and several publications dealing mainly with the etiology of NC should be mentioned: Jaccottet (1959), Heidbreder et al. (1974), Boeckle and Sarre (1976), Massry (1979), and Anderson (1982).

Basic diseases accompanied by hypercalcemia are compared to those which might develop NC without hypercalcemia. The etiological factors presented in

Table 2. Etiology of primary (metastatic) and secondary (dystrophic) nephrocalcinosis

With hypercalcemia	Without hypercalcemia
Hyperparathyroidism	Renal tubular acidosis (complete, incomplete)
- Primary (parathyroid hyperplasia, adenoma, carcinoma)	Hypochloremic alkalosis
	Fanconi syndrome
- Secondary (for example end-stage kidneys)	Renal cell carcinoma
- Tertiary	Tubulointerstitial diseases/lesions
Pseudohyperparathyroidism (tumors)	- Pyelonephritis ± lithiasis
Bone tumors	- Reflux nephropathy
- Primary malignant bone tumors	- Hydronephrosis
- Skeletal metastases	- Cystic kidney
Multiple myeloma	- Medullary sponge kidney
Leukemia	- Analgesic nephropathy
Paget's disease	- Malakoplakia
Sarcoidosis (Boeck's disease)	- Papillary necrosis
Hyperthyroidism	- Mercurial poisoning
Fractures	- Hypoxic lesions
Immobilization	End-stage kidneys
Progressive osteoporosis	Hypercortisolism
Massive idiopathic osteolysis	- Cushing's disease
(Gorham-Stout disease)	- Systemic cortisone therapy
Idiopathic hypercalcemia in childhood	Diuretic therapy
Milk alkali syndrome (Burnett)	Osteoporosis
	Primary and secondary oxalosis
	Magnesium deficiency
	Idiopathic hypercalciuria
	Amyloidosis
	Vitamin A deficiency

Table 2 obviously differ in their incidence and significance for the disease. In primary hyperparathyroidism NC can be an important finding with regard to the course of the disease, whereas the smallest dystrophic calcifications in pyelonephritis are negligible.

Varying conceptions are given on the interrelation between secondary hyperparathyroidism and NC: in accordance with other authors (Jaccottet 1959; Gerok 1976; Anderson 1982) I think that secondary hyperparathyroidism can cause NC; this contradicts the opinion of Reubi (1970).

Morphology of Nephrocalcinosis

Light and electron microscopic findings are presented in order of etiology as listed in Table 2. In particular, primary diseases in which I myself have examined biopsy, nephrectomy, or autopsy material are considered.

Analyzing the literature, a publication by Jaccottet (1959) has to be cited; this author evaluated his own as well as other material for the predisposition of nephron segments to calcification. Figure 2 shows schematic graphs taken from this paper presented side by side for an easier comparison. They demonstrate not only the topography of the calcification but also its extent. Table 3 illustrates the schematic pictures as far as the quantity of the calcification is concerned.

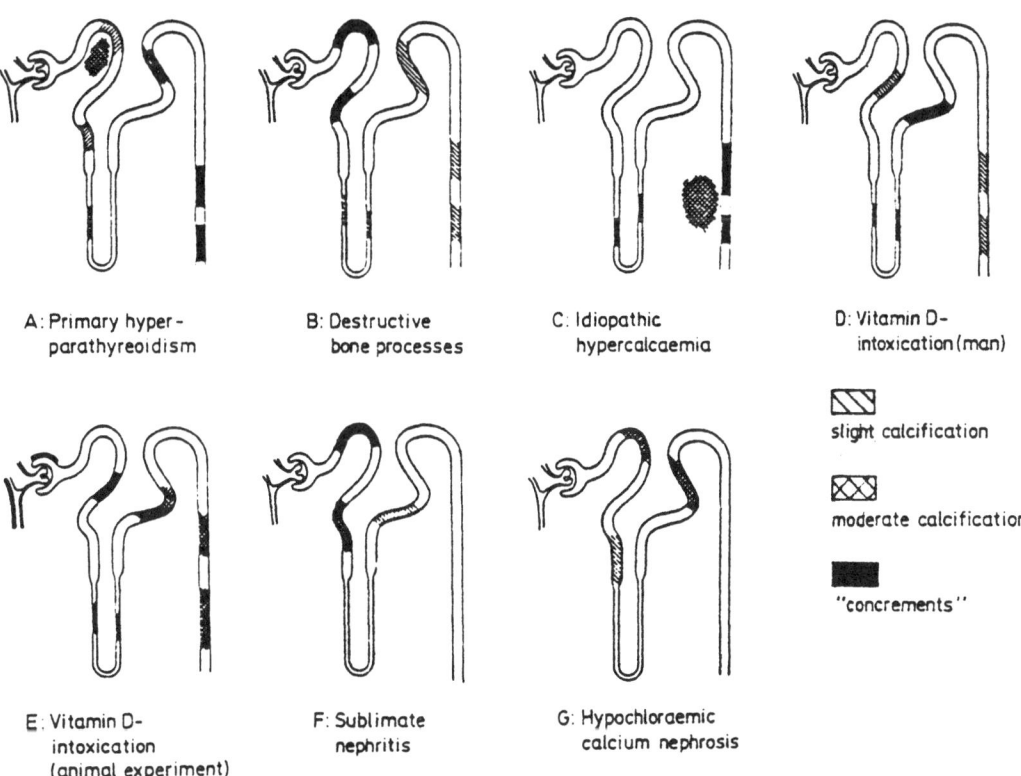

A: Primary hyper-
   parathyreoidism

B: Destructive
   bone processes

C: Idiopathic
   hypercalcaemia

D: Vitamin D-
   intoxication (man)

E: Vitamin D-
   intoxication
   (animal experiment)

F: Sublimate
   nephritis

G: Hypochloraemic
   calcium nephrosis

slight calcification

moderate calcification

"concrements"

Fig. 2. Schematic representation of the nephron segments (*A–G*) affected by nephrocalcinosis in different primary diseases. (Taken from Jaccottet 1959)

Table 3. Distribution pattern of nephrocalcinosis in different primary diseases (Jaccottet 1959)

Primary disease	Glomeruli	Tubular apparatus						Inter-stitium	Arteries
		Proximal tubules		Henle's loop (thin)	Distal tubules		Collecting ducts		
		Convoluted	Straight		Straight	Convoluted			
A. Primary HPT	Rare!	Epith. + BM ±	–	C (+)	Epith. ++	–	C +++	Renal cortex ++	–
B. Destructive bone processes	–	C +++ Epith. +++ BM +++	–	C + Epith. +++ BM +	–	C + Epith. + BM +	–	–	–
C. Idiopathic hypercalcemia	Bowman's capsule ±	–	–	C +++	–	–	C +++	Medulla ++	Media ±
D. Vitamin D intoxication	–	C + Epith. +	–	C + Epith. +	–	C ++ Epith. ++	C + Epith. +	–	–
E. (no consideration)									
F. Sublimate nephrosis	–	Epith. +++ (necroses)	Epith. ++	–	–	Epith. +	–	–	–
G. Hypochloremic calcium nephrosis									
Acute	–	Epith. ++	Epith. +	–	–	Epith. ++	–	–	–
Chronic	–	–	–	Epith. ++	–	Epith. ++ (M.d.)	–	–	–

Abbreviations: HPT, hyperparathyroidism; BM, basement membrane; C, "concrements" (lumina); Epith, epithelium; M.d., Macula densa.

Table 4. Distribution pattern of nephrocalcinosis in my own histologic material

Primary disease	Glomeruli	Tubular apparatus					Collecting ducts	Interstitium	Arteries
		Proximal tubules		Henle's loop (thin)	Distal tubules				
		Convoluted	Straight		Straight	Convoluted			
Primary HPT (n=2)	((+))	Epith.++	Epith.(+)	Epith.(+)	Epith.(+)	Epith.++	Epith.(+)	Cortex, medulla (+)? papilla	–
Pseudo-HPT (n=1)	?	?	?	Epith.(+)	?	?	Epith.(+)	Medulla +	–
Massive osteolysis Gorham-Stout (n=1)	–	–	–	Epith.++	–	Only Macula densa	Epith.++ C+	Papilla +	–
Chronic PN ± lithiasis (n=10)	–	Patchy/dystrophic+	Patchy/dystrophic+	Patchy/dystrophic+	Patchy/dystrophic+	Patchy/dystrophic+	Patchy/dystrophic+	Patchy dystrophic +/+++/+++	–
Reflux nephropathy (n=55)	–	Patchy/dystrophic+	Patchy/dystrophic+	Patchy/dystrophic+	Patchy/dystrophic+	Patchy/dystrophic+	Patchy/dystrophic+	Patchy dystrophic +/+++/+++	–
Cystic kidneys (n=10)	–	Calcification in secretion in cysts			Calcification in secretion in cysts		(+)	Cortex, medulla, papilla (+)	–
End-stage kidneys (n=14)	((+))	Epith. and BM (+)	–	–	–	Epith. and BM ++/+++	–	Medulla, papillae patchy dystrophic	Elastica interna –

Abbreviations: HPT, hyperparathyroidism; BM, basment membrane; C, "concrements" (lumina); Epith., epithelium; PN, pyelonephritis.

Ahead of the discussion, in Table 4 the semiquantitative findings of my own cases allow direct comparison with Table 3, as far as identical primary diseases are concerned.

Primary Hyperparathyroidism

It was striking that two of my own kidney biopsy cases of primary hyperparathyroidism were accompanied by a transitory acute renal failure at the time of diagnostic biopsy. NC affected mainly the proximal and distal convoluted tubules (cf. Table 4). In contrast to the findings of Jaccottet (1959) I found no extensive calcification in the area of the collecting ducts (cf. Fig. 2; Table 3). The identification of calcium salts can barely be achieved by the PAS reaction and so it is restricted to larger calcium deposits (Fig. 3); the Kossa reaction, however, allows an exact qualitative and quantitative determination (Fig. 4). Epithelial cells of the proximal and distal convoluted tubules are completely calcified. In the case of distinct calcification the determination of the nephron segments affected was barely or no longer possible (cf. Jaccottet 1959). Apart from calcifications which respect the borders of the renal tubules and which can – at least from the view of their development – be regarded as metastatic NC, this figure also shows dystrophic calcification adjacent to destroyed tubules (Fig. 4).

As demonstrated by light microscopy, the different nephron sections and even individual tubular epithelial cells neighboring each other exhibit different quantities of calcification. Electron microscopic examinations confirm this: thus sometimes tubular epithelia with only fine granular calcification of mitochondria can

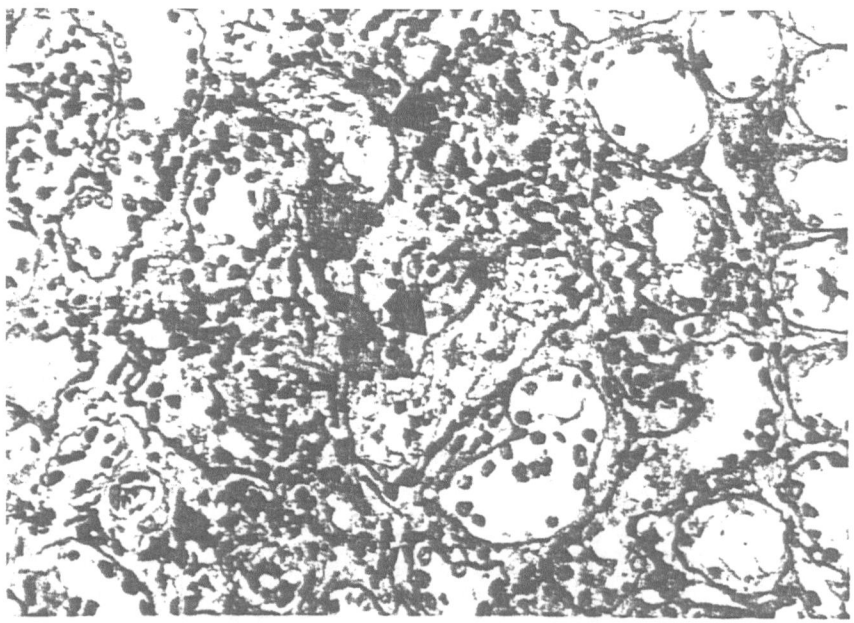

Fig. 3. Primary hyperparathyroidism: two calcified cortical tubules (*arrows*) which are difficult to identify (proximal tubules?), PAS reaction. ×313

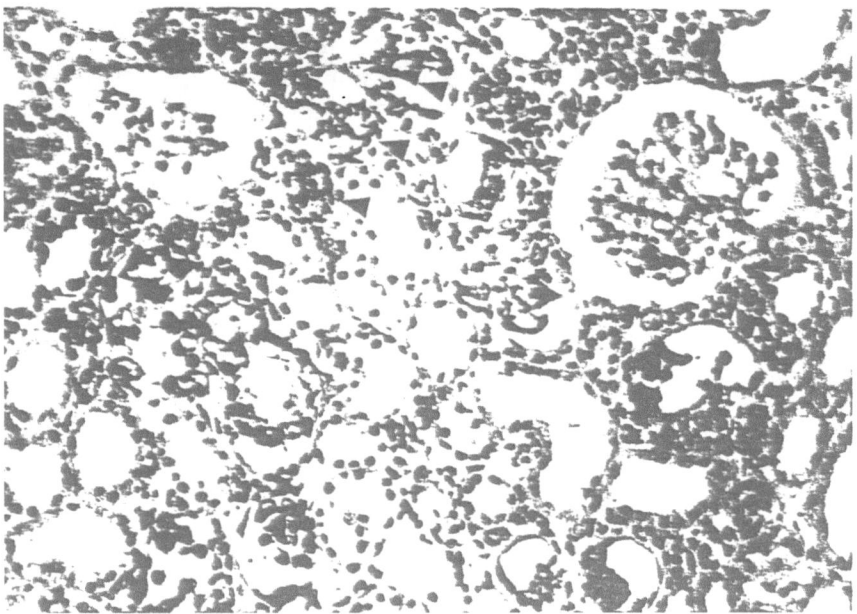

Fig. 4. Primary hyperparathyroidism: distinct calcification of the epithelial cells of the proximal convoluted tubules, beside them a plaque-shaped calcification at the margin of an area of cell destruction (*arrows*), × 200

be found (Fig. 5), while other epithelia contain larger, partially membranelike bordered (Fig. 6), partially irregular calcifications in the cytoplasm (Fig. 7). Analogous to tubular epithelia appearing strongly calcified in light microscopy following the Kossa reaction, the cytoplasmic calcification is more distinct electron microscopically (Fig. 8). On the base of a tubular epithelial cell I observed a displacement of the basal labyrinth in an apical direction by a plaquelike calcium deposit which was located between the basal labyrinth and the noncalcified and still preserved tubular basement membrane (Fig. 9).

A preferred deposition of calcium salts in the outer medullary zone was described in primary hyperparathyroidism in man (Cottier 1980). In less severe cases, Henle's loops, the distal convoluted tubules and the collecting ducts are supposed to be affected. Other authors, on the contrary, regarded this area of the papilla as being the most predisposed to NC in primary hyperparathyroidism (Haas and Dambacher 1982).

Most likely to be interpreted as complications are the alterations described by Heptinstall (1983), namely an alteration of normal parenchyma and sectorlike scars the development of which might be attributed to an obstruction of bigger collecting ducts. This author found calcification of the tubular basement membranes not only within the scars but also outside of the scars. Each nephron segment can principally yield calcified tubular epithelia (Heptinstall 1983).

Animal experiments produced evidence most similar to that in man. In dogs treated with extracts of parathormone in NC Carone et al. (1960) found structural alterations mainly in the ascending limb of Henle's loop, in the distal convoluted tubules, and in the collecting ducts of cortex and medulla. In addition, they described focal degeneration of epithelia with calcification, necrosis and obstructing

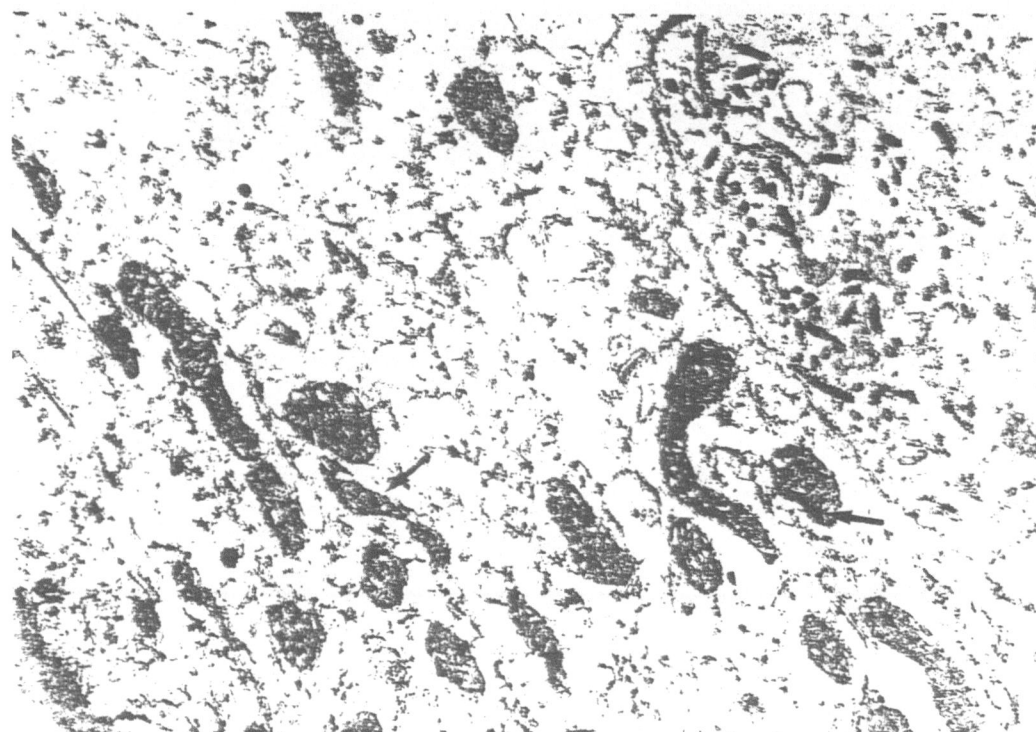

Fig. 5. Primary hyperparathyroidism: discrete calcification in the form of tiny calcium granules (*arrows*) in structurally preserved mitochondria. × 24,600

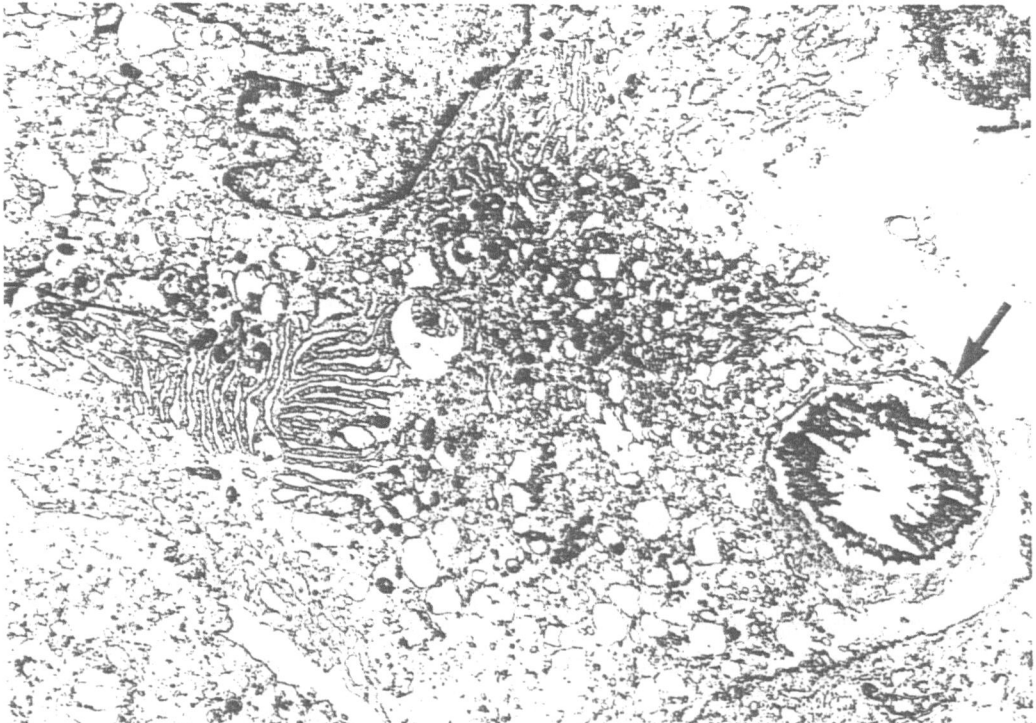

Fig. 6. Primary hyperparathyroidism: plaque-shaped, membrane-like bordered calcium deposits in the cytoplasm of a tubular epithelial cell (*arrow*). × 9400

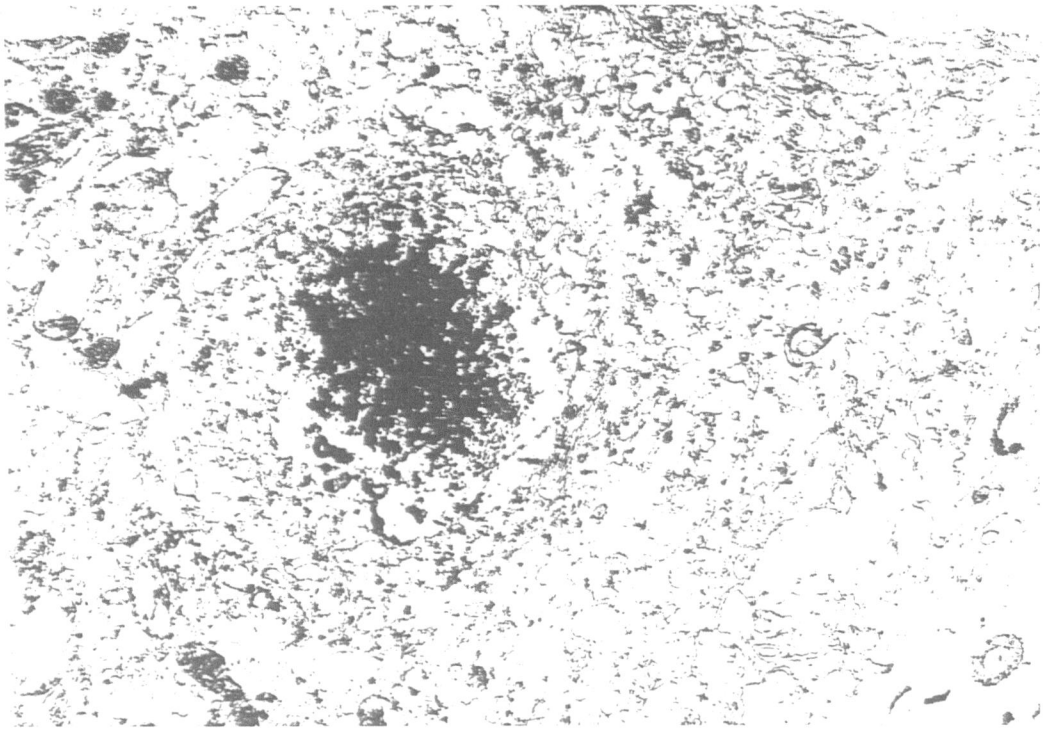

Fig. 7. Primary hyperparathyroidism: cytoplasm with black irregular calcium deposits. × 19,600

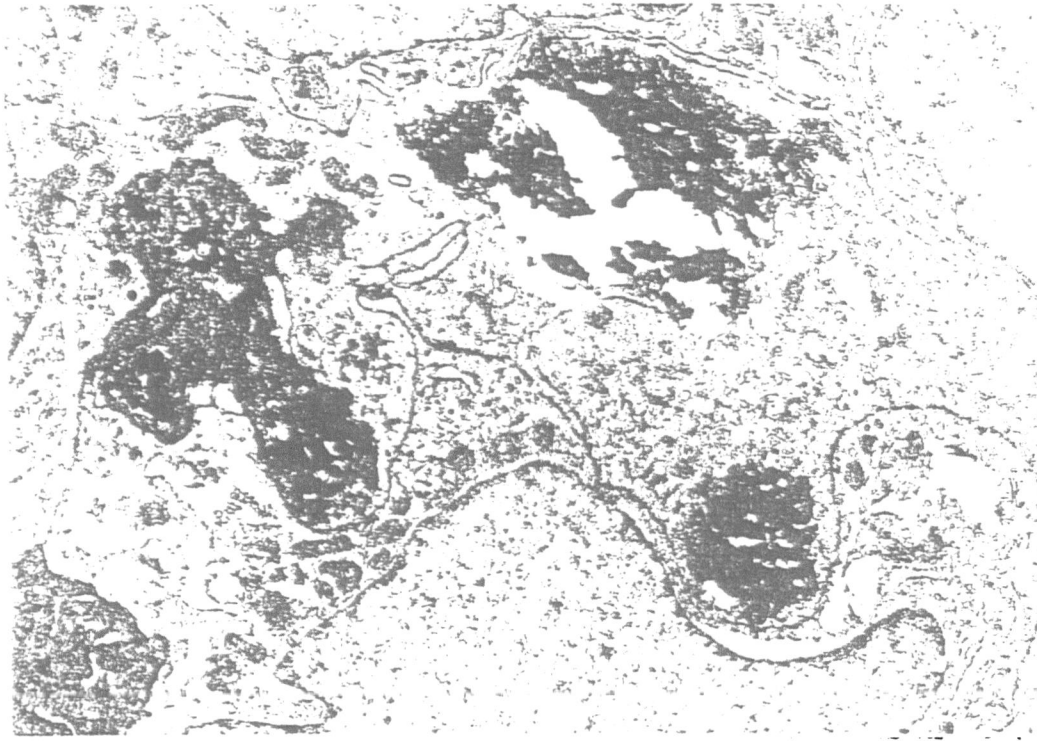

Fig. 8. Primary hyperparathyroidism: larger partially plaque-shaped and dispersed calcium deposits in the cytoplasm of several adjacent tubular epithelial cells. × 15,400

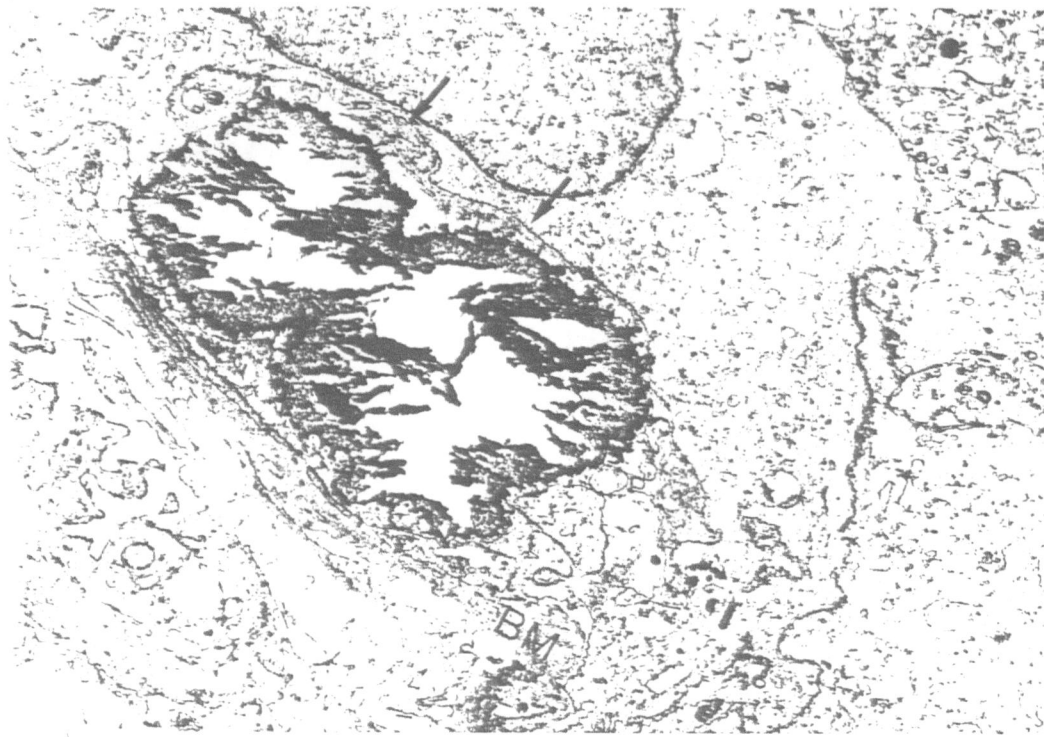

Fig. 9. Primary hyperparathyroidism: base of a tubular epithelial cell; on the upper side of the basement membrane (*B.M*) a coarse plaque-shaped calcium deposit. At the border to the cell base displaced by this is probably newly formed basement membrane material (*arrows*). × 12.200

casts. Interstitial calcification did not occur to any significant degree, although it may appear in longer lasting experiments (Anderson 1982).

Scarring lesions, for instance scarred glomeruli and smaller cortical cysts as well as round cell infiltrates, are concomitant findings. Then the vessels show severe arteriosclerotic and hypertensive alterations.

The significance of structural changes caused by NC lies in the possibility of renal failure in the course of time (Sanderson 1959; Boeckle and Sarre 1976; Cottier 1980; Heptinstall 1983; Bichler et al. 1985). This occurs mainly with cases of primary NC; secondary NC, in contrast, being mostly focal, does not produce a particular impairment of kidney function (Cottier 1980). The authors mentioned pointed out that there was a reduction of renal concentration ability; the glomerular filtration rate could be decreased and the development of a chronic renal insufficiency was possible. The latter might be due to a reactive cortical interstitial fibrosis resulting from tubular calcification.

Pseudohyperparathyroidism

The production of parathormonelike substances in different tumors outside of the parathyroidea is called pseudohyperparathyroidism (Cottier 1980). The proximal tubules are thought to be primarily affected (Jaccottet 1959).

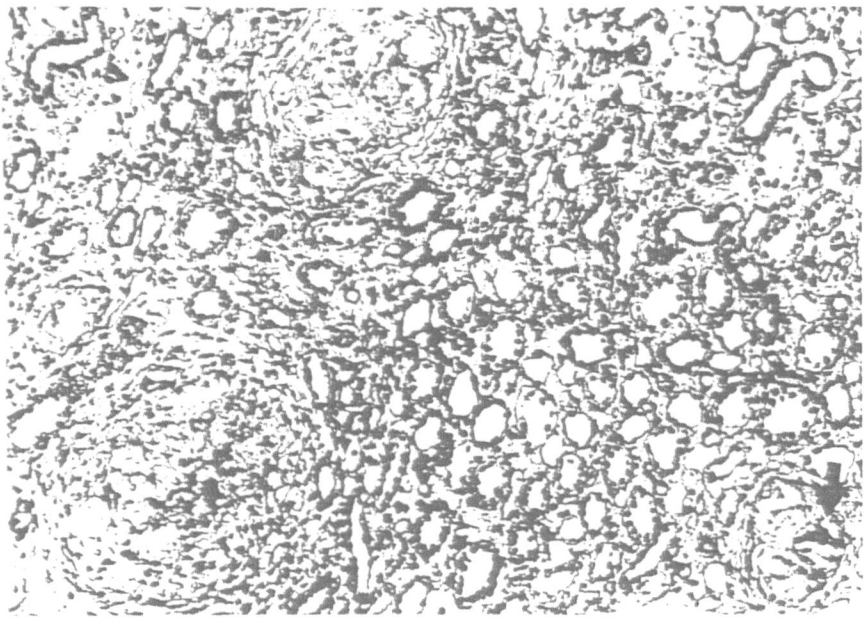

Fig. 10. Pseudohyperparathyroidism: in the renal medulla of a biopsy specimen three granuloma-like structures; in the smallest granuloma is a calcium deposit (*arrow*). Kossa reaction, × 192

In a corresponding biopsy I found granulomalike structures of epithelioid cells and a few multinuclear giant cells in the renal medulla; focally I detected calcium within the granulomas (Fig. 10).

Independent of the underlying disease of NC a granulomatous reaction has rarely been reported, for instance in idiopathic hypercalcemia (Anderson 1982). These findings require the exclusion of other granulomatous diseases, particularly of sarcoidosis or productive renal tuberculosis.

Massive Idiopathic Osteolysis

This disease (synonym: Gorham-Stout disease) is discussed as an example of NC caused by an increased mobilization of calcium from the skeleton. Other diseases with destructive bone processes (Jaccottet 1959) are the multiple myelomas, sarcomas, lymphomas, and bone metastases of carcinomas; they can be accompanied by metastatic NC with predominant calcification of the tubular basement membranes, the epithelia, and the lumina of the proximal convoluted tubules (cf. Fig. 2, Table 3).

In my own autopsy case of a 79-year-old man I found an extreme osteolysis which had developed in the lower extremities, the pelvis, and lumbar vertebral column within the last 10 years. The Gorham-Stout disease is defined as hemangiomatosis and/or lymphangiomatosis of the bones with consecutive destruction and fibrosis: parts of both femora and tibiae had almost disappeared in this patient. Both kidneys were of normal size in the autopsy; it could be seen from the medical history that the serum creatinine values had remained within the normal range until the patient's death.

The histological evaluation of calcification in the renal cortex was mostly negative. Only distal nephron parts (exclusively areas of the macula densa) were severely calcified (Fig. 11).

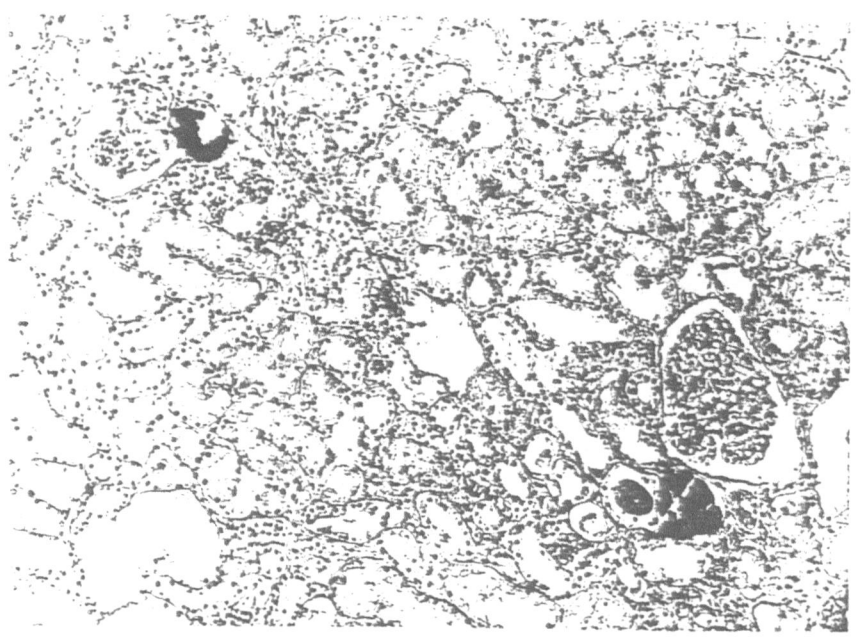

Fig. 11. Massive idiopathic osteolysis: completely calcified macula densa in the cortex of both kidneys (autopsy material). Kossa reaction. × 90

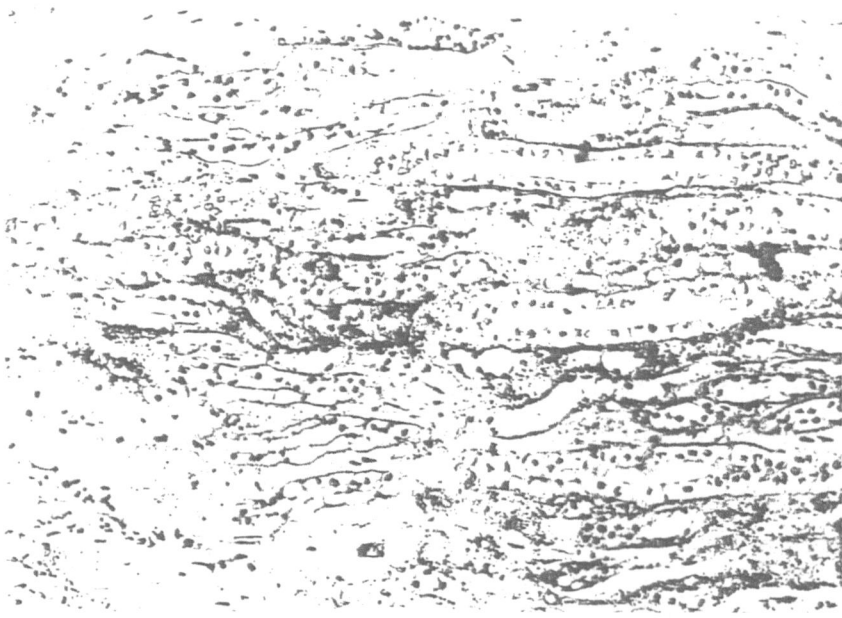

Fig. 12. Massive idiopathic osteolysis: in the renal medulla finely granular interstitial calcification of the matrix. Kossa reaction. × 128

In the renal medulla a rather distinct fine granular interstitial calcification of the matrix could be demonstrated (Fig. 12), while calcium deposition in the epithelia of the thin limbs of Henle's loops showed no considerable calcification (Fig. 13). Isolated thick limbs of Henle's loops were severely calcified; however, some renal tubules were found which had only a few completely calcified epithe-

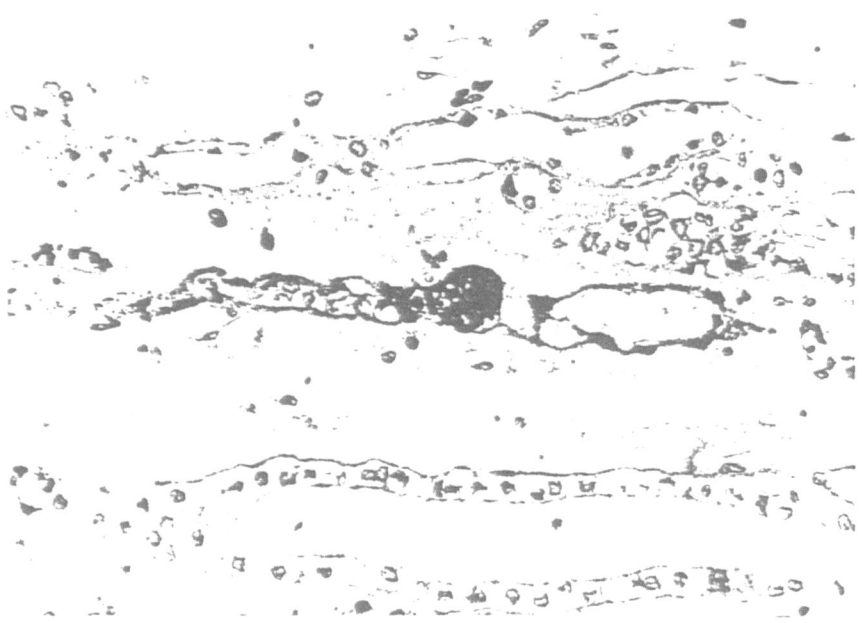

Fig. 13. Massive idiopathic osteolysis: in addition distinct medullary calcification in the epithelial cells of Henle's loops. Kossa reaction, × 250

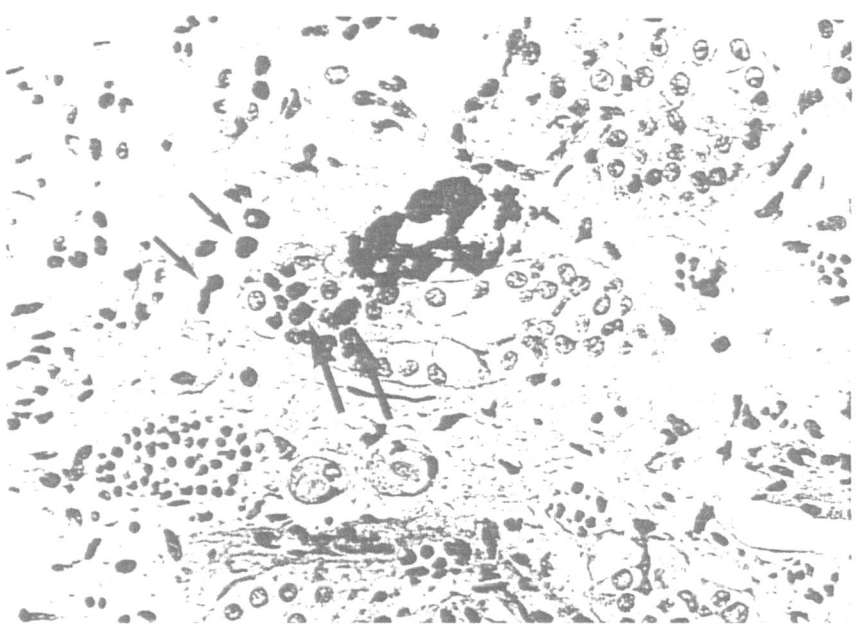

Fig. 14. Massive idiopathic osteolysis: in the renal medulla are completely calcified tubular epithelial cells, and a few partly calcified epithelial cells (*long arrows*); in addition, two interstitially localized, calcified cells (*short arrows*). Kossa reaction, × 370

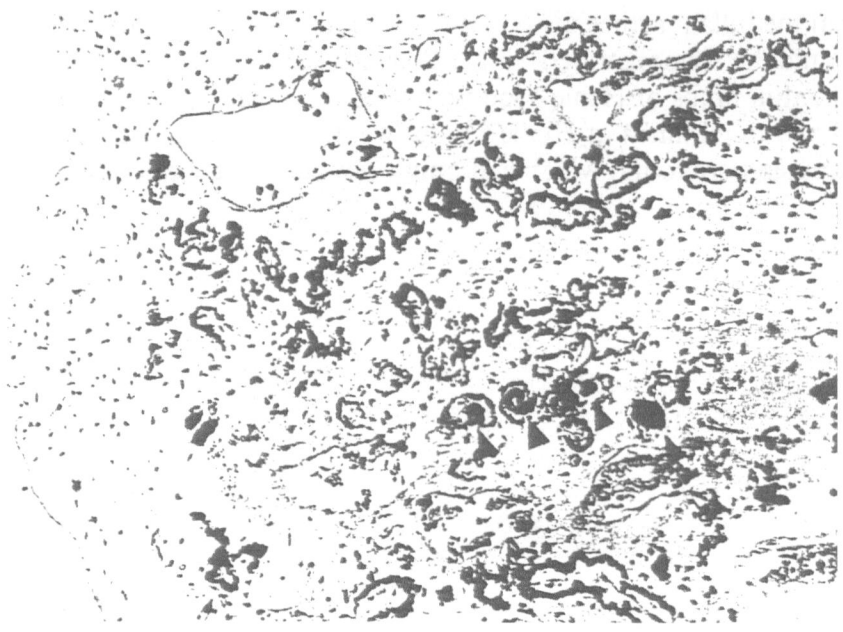

Fig. 15. Massive idiopathic osteolysis: part of a renal papilla with severe calcification of the epithelial cells of the collecting ducts. Several tiny concrements in the lumen of calcified collecting ducts (*arrows*). Kossa reaction. × 100

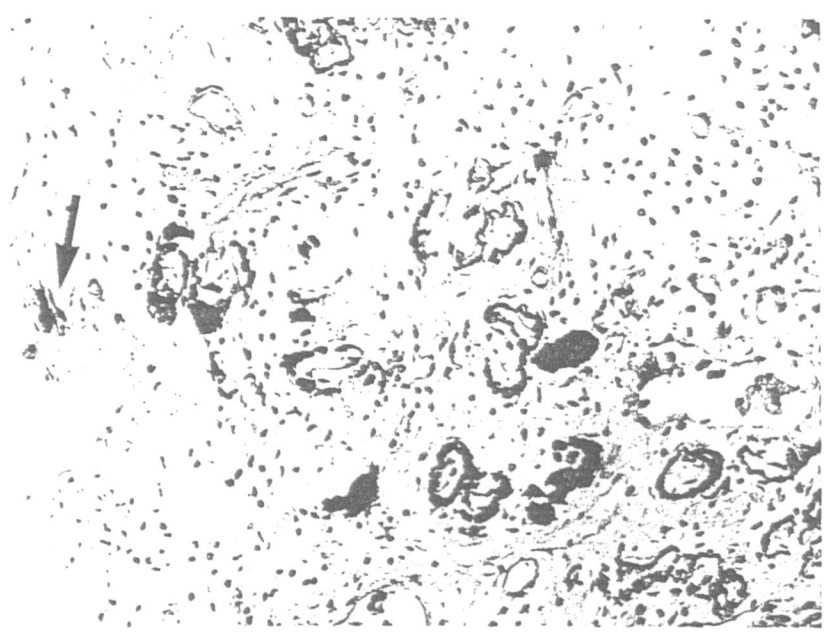

Fig. 16. Massive idiopathic osteolysis: in another papilla similar conditions, additionally just below the papillary surface a small calcification of the type of Randall's plaques. Kossa reaction. × 128

lia; interstitial cells which could not be clearly characterized were also severely calcified (Fig. 14). Comparing renal cortex, medulla, and papillae a gradation of NC was seen, i.e., the renal cortex was slightly, the medulla moderately, and the papillae distinctly calcified (cf. Table 4). Not only were the epithelia of the collecting ducts severely calcified but concentric calcifications were also found in the presence of microliths in the lumina of similarly altered collecting ducts (Fig. 15).

Smaller calcifications which could be identified just below the papillary surface are similar to Randall's plaques (Fig. 16); they will be discussed later on. Although the Gorham-Stout disease corresponds to a destructive bone process, these findings do not conform with the simplified conceptions given by Jaccottet (1959). The attempt to classify a certain pattern of NC into an entity or into a group of basic conditions obviously fails in this particular case.

Idiopathic Hypercalcemia

According to Jaccottet (1959) particularly distinct calcifications are supposed to occur in the thin segments of Henle's loop and in the area of the collecting ducts, in addition to interstitial calcification in the medulla (cf. Fig. 2, Table 3). The predisposition of Henle's loop – as mentioned before – accords with the statement that the renal medulla is the most affected area if light microscopy is done with low magnification (Anderson 1982).

Sarcoidosis (Boeck's Disease)

About 20% of sarcoidosis cases are thought to be accompanied by hypercalcemia. If NC appears, the epithelia of the collecting ducts are the most affected; in addition calcium casts are found in the collecting duct lumina and eventually interstitial calcification can be seen (Klatskin and Gordon 1953; Anderson 1982). Up to now it is still unclear to what extent NC and/or granulomas are responsible for the possibility of renal insufficiency (Scholz and Keating 1956).

Vitamin D Intoxication

There are few reports on the renal conditions in vitamin D intoxication in man (Jaccottet 1959; Anderson 1982). The distal convoluted tubules are thought to be more affected by NC than the proximal convoluted tubules, the thin segments of Henle's loop, and the collecting ducts (cf. Fig. 2, Table 3). In addition to a calcification of the epithelium, "concrements" can be found in all these sites as well; the term "concrement" should be replaced by "cast" or "microlith". The conditions in animal and man seem to differ: in animal experiments the alterations progress with time, starting from the proximal tubular epithelia but affecting the tubular basement membranes and arterial vessel walls later (Jaccottet 1959).

Other authors have described calcification of the tubular basement membranes and the contents of the tubular lumina in vitamin D intoxication (Seifert et al. 1975). According to Anderson (1982) all segments of the nephron may be affected.

Detailed descriptions on the site of deposition of calcium salts in vitamin D intoxication can be taken from animal experiments which used electron microscopic studies (Engfeldt et al. 1962; Gerlach and Themann 1965; Scarpelli 1965; Schäfer 1979). Summarizing these publications NC is of the metastatic type with a special calcium phosphate deposition. Due to the calcium, the proximal tubular epithelia show structurally damaged mitochondria, a distention of the endoplasmatic reticulum and – after a release of apical calcium-containing cytoplasm – casts in the tubular lumina. As mentioned in the description on massive idiopathic osteolysis in these animal experiments individual tubular epithelia remain unchanged, whereas others show damaged mitochondria exclusively.

Mostly granular calcium deposits have been described for the basement membrane of the proximal tubules. Scarpelli (1965) pointed out that there is an accentuation of the corticomedullary junction; depending on the severity of the vitamin D intoxication, at least interstitial calcification can occur.

Although the main localization of NC is undoubtedly in the proximal convoluted tubules, considerably altered collecting ducts have also been described (Engfeldt et al. 1962).

Renal Tubular Acidosis

The incomplete form of renal tubular acidosis is much more frequent than the complete form (Marquardt 1973). In this renal disease which is not accompanied by hypercalcemia and which was presented schematically by Jaccottet (1959) as hypochloremic calcium nephrosis (cf. Fig. 2, Table 3), the calcification in the proximal tubules is said to be slightly heavier than in the straight distal tubules. The tubular epithelia are particularly calcified with a striking predisposition exhibited by the macula densa.

Anderson (1982) distinguished between juvenile and adult forms. In the juvenile form NC affects the medulla and pyramidal base and the collecting ducts show calcification of the epithelia and the basement membrane. The lumina of collecting ducts contain microliths. Interstitial calcification can be accompanied by a giant cell reaction. In the adult form there are cortical scars with different causes, e.g., tubular obstruction, infections, nephrolithiasis. Renal tubular acidosis involves metastatic NC at least in the early phase of the disease.

Renal Cell Carcinoma

Another example of primary diseases without hypercalcemia accompanied by NC (in the broader sense) are renal cell carcinomas (formerly: hypernephroma, hypernephroid carcinoma). Calcification is not rare and its manifestation as a psammo-

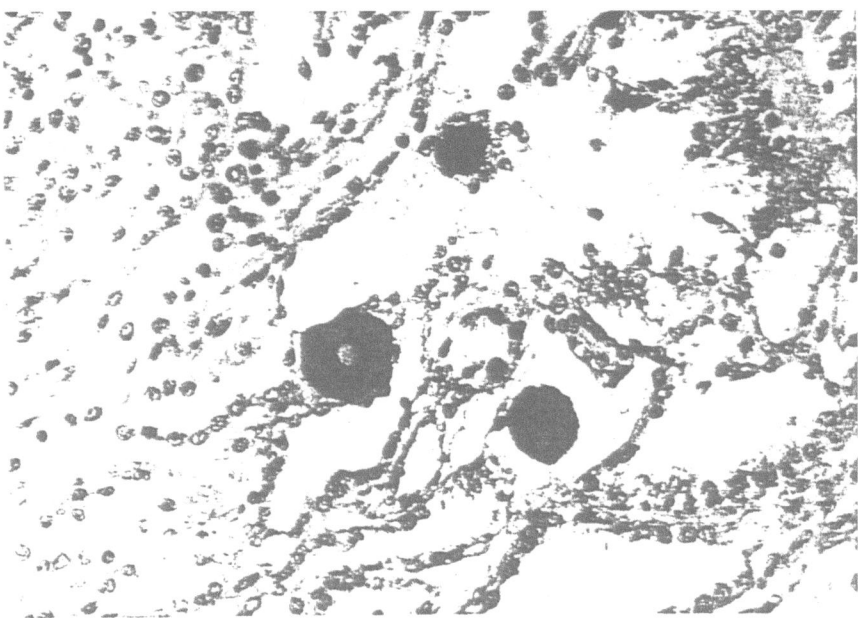

Fig. 17. Renal cell carcinoma: three psammomalike partially stratfied calcifications surrounded concentrically by tumor cells. Kossa reaction. × 250

malike structure might be due to a calcification of a tumor secretion product (Fig. 17). However dystrophic calcification of spontaneous, regressively altered renal cell carcinomas seem to be rather rare findings.

Chronic Pyelonephritis Without Nephrolithiasis

A random sample of five cases of chronic nonspecific pyelonephritis without nephrolithiasis exhibited very heterogeneous pictures in the Kossa reaction. The spectrum ranged from ultrafine, patchy dystrophic calcification, mainly in the medullary and papillary region, less frequently in the cortex (compare Anderson 1982), to extended coarse, plaque-shaped calcifications exclusively in the area of the medulla and papilla (Fig. 18). No close correlation could be found between calcification and the extent of the tubular destruction.

While Fig. 18 a gives the impression of a homogeneous deposition of calcium salt, it is evident on polarization that not only calcium phosphate and/or carbonate but also calcium oxalate, which is birefringent in polarized light, can be deposited (Fig. 18 b). The pathologic value of a moderately distinct NC in chronic pyelonephritis can be neglected.

Investigating 168 nephrectomy and autopsy kidneys (primary disease: tuberculosis, hydronephrosis, pyelonephritis, nephrolithiasis), Anderson and McDonald (1946) found small "calculi" in the medullary rays and in addition, in 50% of the kidneys, plaques could already be seen in the first histological section. This is more frequent than the finding of nephrolithiasis.

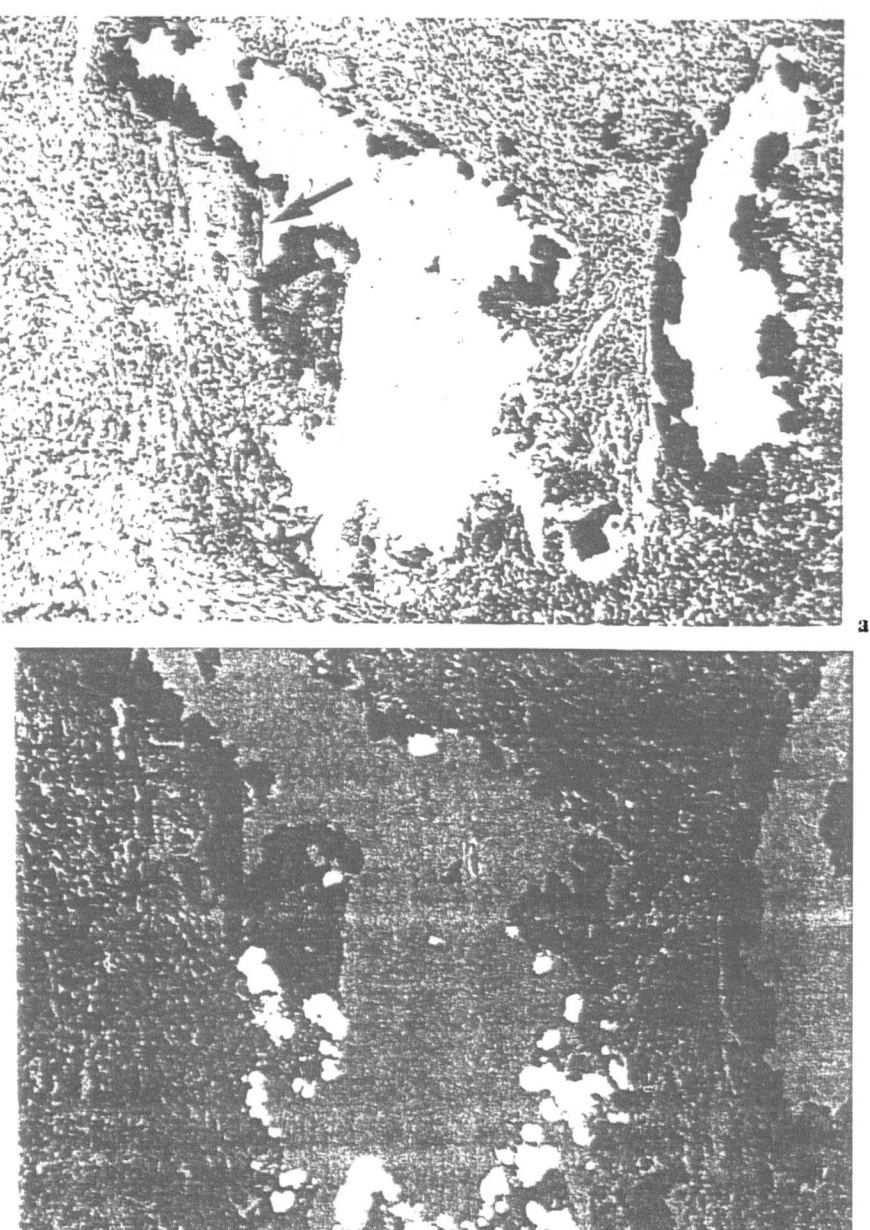

Fig. 18 a, b. Chronic pyelonephritis without nephrolithiasis. **a** Coarse plaque-shaped calcification in the renal medulla with complete destruction of the tubular apparatus and dense round cell infiltration. Multinuclear giant cell (*arrow*). **b** Same section in polarized light: birefringent calcium oxalate (*white*), calcium phosphate and or carbonate (*black*). Both figures: Kossa reaction. × 80

Chronic Pyelonephritis with Nephrolithiasis

Of five cases of nephrectomy which were examined for calcium salt deposits, none demonstrated a more distinct NC with existing nephrolithiasis than the previous group. There was also a wide range of calcification which was almost exclusively

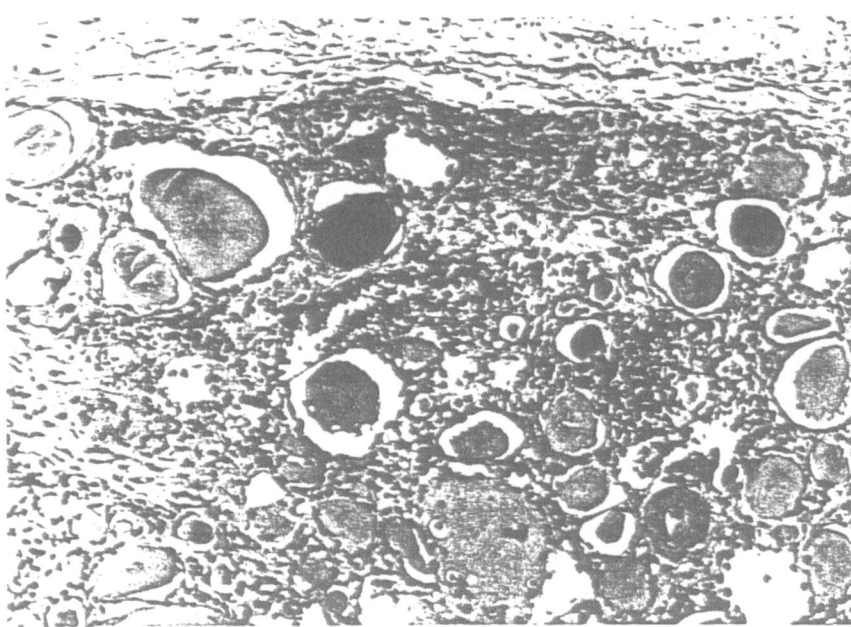

Fig. 19. Chronic pyelonephritis in nephrolithiasis: subcapsular cortical area with distinct formation of thyroidlike lesions. Three follicles show a calcification of the light gray secretion product. Kossa reaction, × 160

dystrophic calcification caused by cell death. An exception was a case where calcification of inspissated secretion occurred within the thyroidlike lesions (Fig. 19). In the majority of pyelonephritic kidneys with thyroidlike lesions, which can subsist almost unchanged for many years and decades, the secretion product in the lumen of the follicles demonstrated no tendency towards calcification. This type of calcification is hard to define: it may be thought of as metastatic NC or as dystrophic NC, depending on whether the secretion product in the follicles is regarded as a pathologic deposition or not.

Reflux Nephropathy

The systematic exploration of 55 cases of reflux nephropathy (a special type of pyelonephritis) demonstrated similar findings as in the two preceding groups (compare Laberke 1987). Medulla and papillae were predisposed to a dystrophic NC (Fig. 20). Dystrophic collecting ducts were found in the area of the medulla as an indication of an already intrauterine onset of damage of the parenchyma, some of them containing psammomalike concrements (Fig. 21). Sometimes the high epithelium of these dysplastic collecting ducts was calcified as well.

As a rule, the cortical parenchyma was the least affected by calcification in reflux kidneys; in one case, however, I saw not only single or several completely calcified tubular epithelia but also glomeruli with distinct calcification of the cells which were most likely podocytes (Fig. 22).

H. G. Laberke

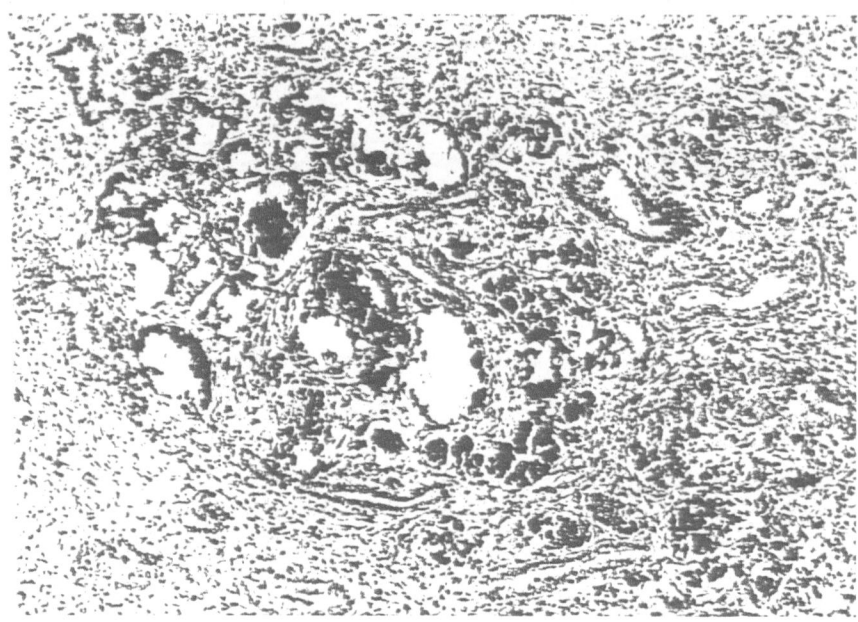

Fig. 20. Reflux nephropathy: in the medulla considerable destruction of the tubular apparatus with plaque-shaped calcification (dystrophic nephrocalcinosis). Increase in connective tissue and round cell infiltration. Kossa reaction, × 80

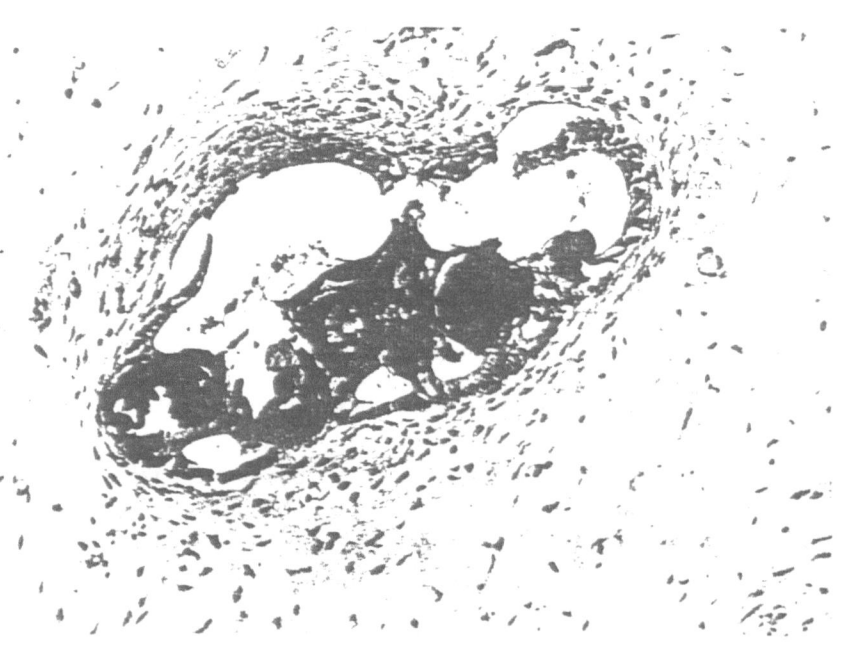

Fig. 21. Reflux nephropathy: dysplastic collecting duct in the medulla with concentrically layered fibromuscular tissue. Epithelium and lumen partially calcified with psammomalike appearance (*black spheres*). Kossa reaction. × 192

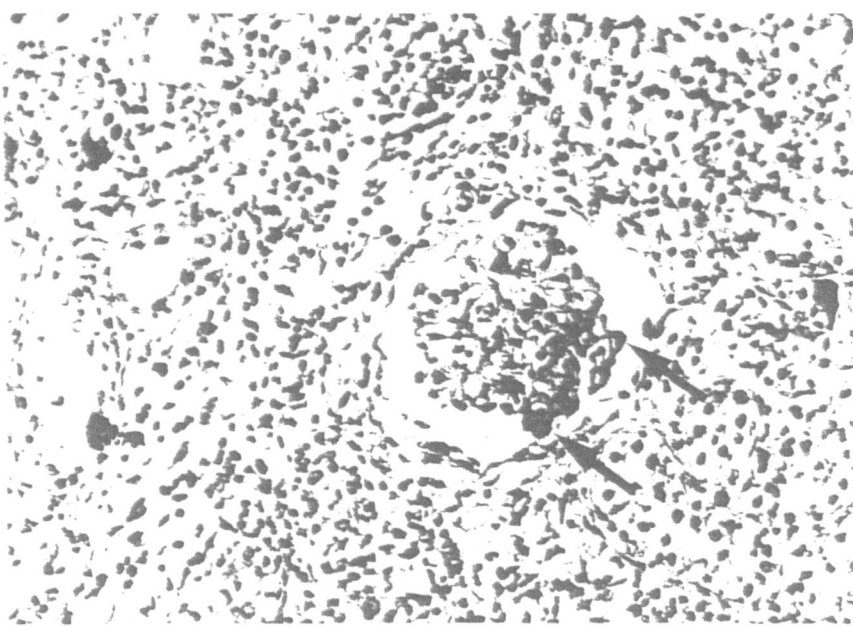

Fig. 22. Reflux nephropathy: cortical parenchyma with impressive tubular destruction. Single tubular epithelial cells are severely calcified (*left side*). In addition remarkable calcification of glomerular cells, most likely podocytes, is shown (*arrows*). Kossa reaction, × 250

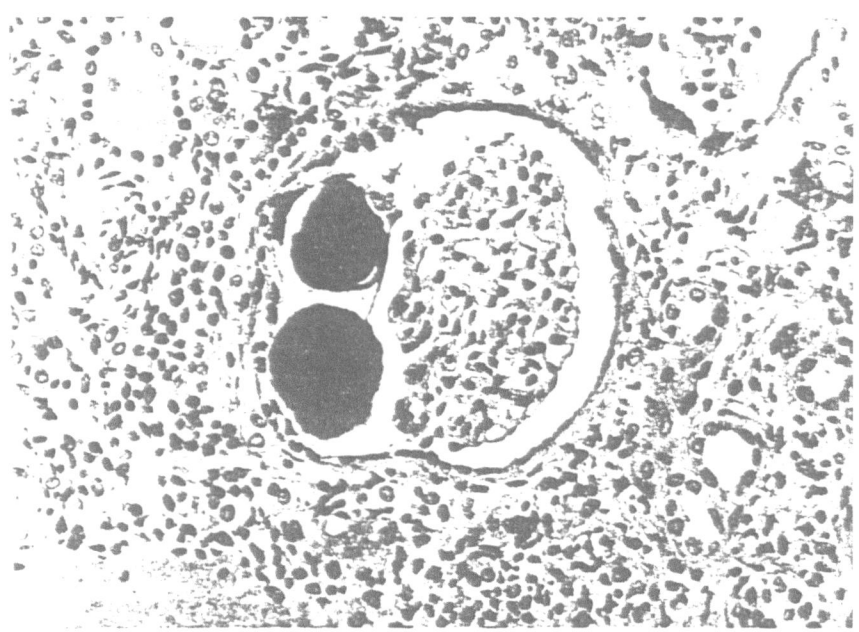

Fig. 23. Reflux nephropathy: glomerulus with unusual psammomalike calcification within Bowman's space. Kossa reaction, × 300

Glomerular calcium deposits had been observed by Jaccottet (1959) only sporadically in the case of primary hyperparathyroidism as a basic condition. They played a minor role in the patients examined and were thus particularly conspicuous when psammomalike structures could be seen in the lumen of Bowman's capsule as in a further case of reflux nephropathy in childhood (Fig. 23). It is uncertain whether they are inspissations and subsequent calcification of primary urine in Bowman's space or a destruction of single glomerular capillaries.

Renal Tuberculosis

It is well-known that caseous organic or lymph node tuberculosis has a tendency towards secondary calcification. For a nephrectomy case in a state after caseous-cavernous renal tuberculosis, calcified and ossified areas were found in scarred connective tissue (Fig. 24); ossification can also appear in other organs under similar conditions.

Hydronephrosis

The situation in kidneys with urinary tract obstruction is probably similar to other diseases with primary damage of the tubulointerstitial system. In 1972 Weller and coworkers reported on interstitial calcification particularly in the area of

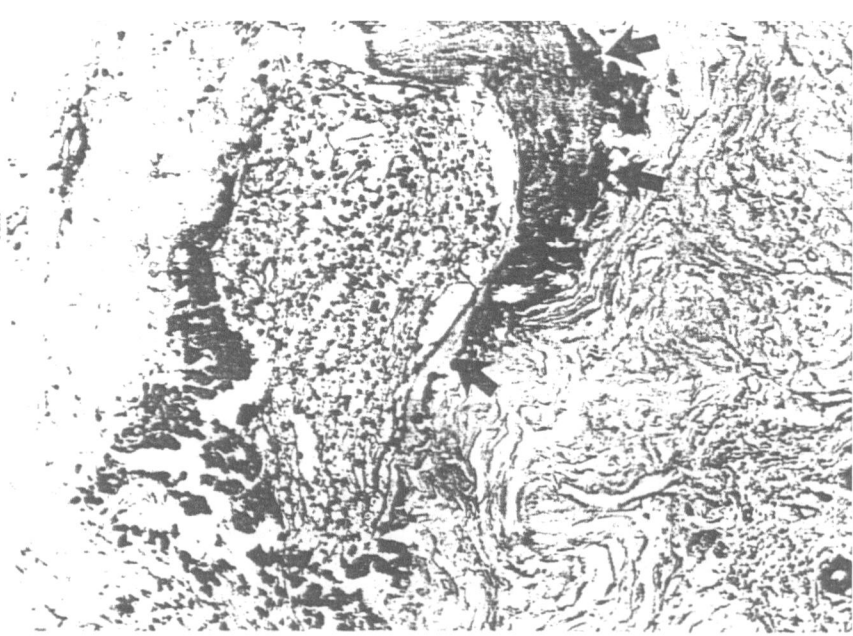

Fig. 24. Renal tuberculosis: scarred renal tuberculosis with plaque-shaped calcification and circumscribed ossification (*arrows*). PAS reaction. × 128

Henle's loop in three normocalcemic cases of hydronephrosis. Electron microscopically, they saw irregular calcium deposits in the papillae, partially at the tubular basement membrane of the thick Henle's loop, and also in the basement membrane of smaller vessels. The basement membrane of collecting ducts, however, was mostly free of calcium deposits.

Cystic Kidneys

The kidneys of 10 patients with congenital cystic kidneys showed no or only slight calcifications. Besides a discrete calcification of the secretion in the cystic lumina there was patchy dystrophic calcification (Fig. 25).

Analgesic Nephropathy

In the early stage of analgesic nephropathy, in which light microscopically, no pathological alterations could be found in the renal cortex. incidental findings by electron microscopy showed tiny glomerular calcifications. Two roundish calcifications were localized in the mesangium in a subendothelial position (Fig. 26). Metastatic calcification of the glomerular basement membrane is a very rare finding, there being only one case of multiple myeloma reported in the literature (Ross and Chin 1970).

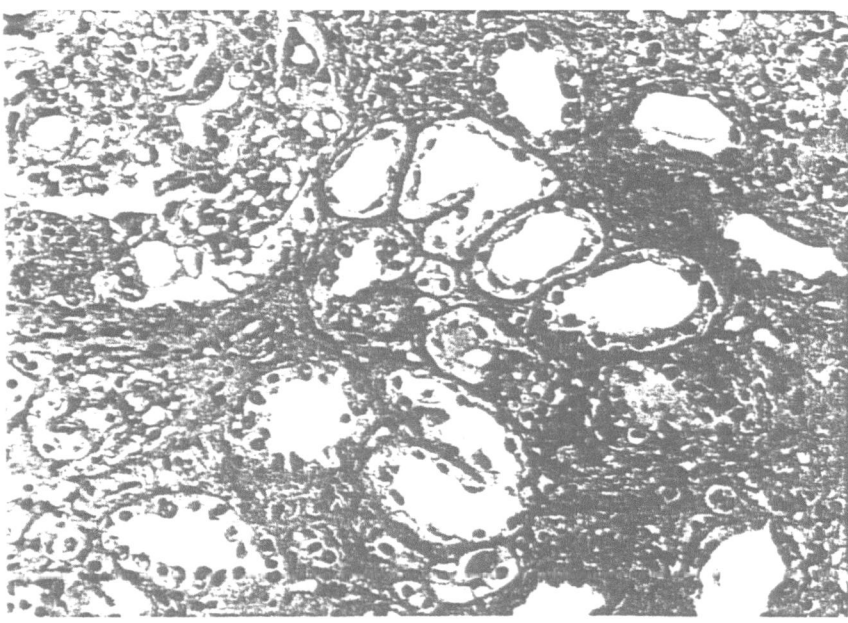

Fig. 25. Cystic kidney: in the renal cortex small plaque-shaped interstitially localized calcifications, most likely due to destroyed renal tubules. Kossa reaction, × 250

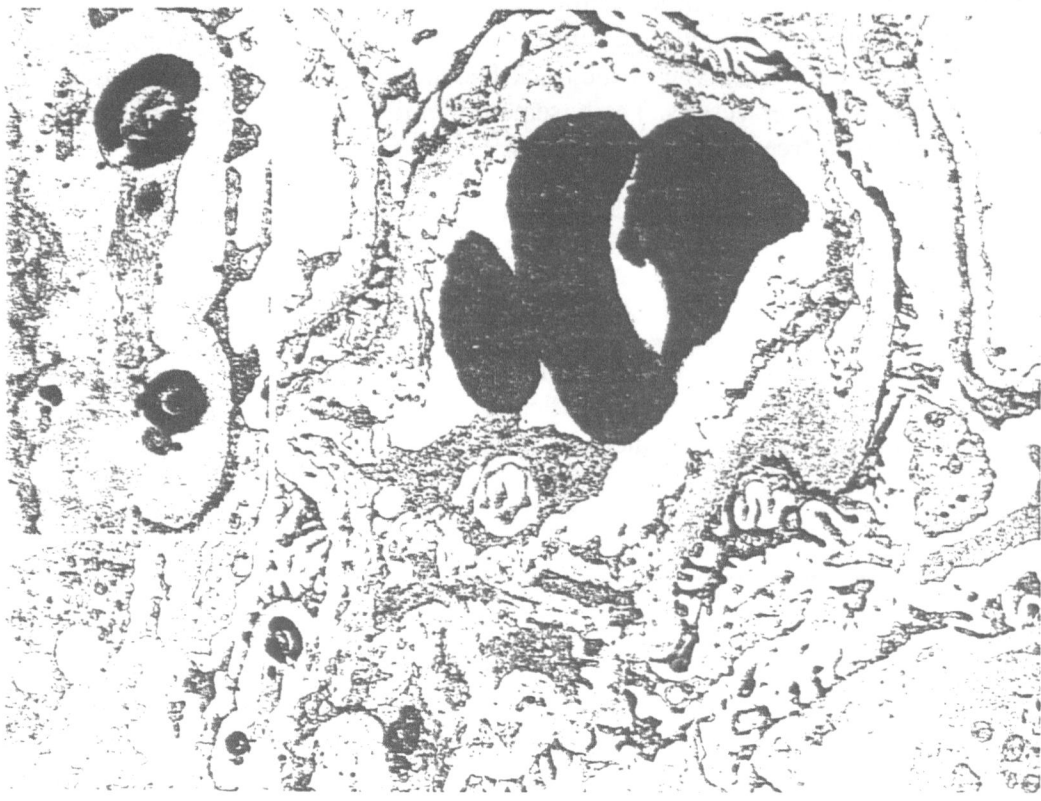

Fig. 26. Early stage of analgesic nephropathy: part of a glomerulus, incidental finding of two spherical subendothelially localized calcifications, × 9400, inset × 24.600

It is doubtful whether the findings presented in Fig. 26 are pathognomonic for analgesic nephropathy. In this disease usually secondary calcified and sometimes also ossified papillary necroses can be seen.

Additional findings of Hennis et al. (1982) should be mentioned, where an acute analgesic nephropathy was induced in the rabbit. Light microscopically the NC started on the corticomedullary junction in the area of the proximal tubules (straight portions) with later heterogeneous effects on the nephrons. They found these alterations in the brush border first, later on intracellularly, and still later intrarenally, but not interstitially. The collecting ducts remained unaffected.

Mercurial Poisoning

In so-called sublimate nephrosis (cf. Fig. 2, Table 3) the severe calcifications of the epithelial cells of the proximal convoluted tubules are of a dystrophic nature, i.e., the result of toxically induced tubular necrosis. Similar alterations of the epithelium in the distal convoluted tubules are less pronounced (Jaccottet 1959). Comparable topographic relations of tubular necrosis induced by sublimate nephrosis were described by Bulger and Dobyan (1984).

End-Stage Kidneys

Fourteen end-stage kidneys which I examined for NC were conspicuous by partly distinct calcification of the renal cortical tubules, starting from the basement membrane and followed by calcification of tubular epithelial cells. The nephron sections could hardly be characterized in these cases since atrophic tubules which cannot be identified clearly seem to be predisposed to calcification. Different sites in the same end-stage kidney (Fig. 27 a–c) show calcification firstly of the tubular basement membrane (Fig. 27 a), then of the basement membrane and basal portions of the tubular epithelium (Fig. 27 b), and finally of all tubular epithelial cells of some nephron segments (Fig. 27 c). While the calcification of the basement membrane can be regarded as a metastatic NC. the more pronounced calcification of the epithelia might be of a dystrophic nature in a later phase.

Isolated scarred and secondarily calcified glomeruli were found in only one end-stage kidney (Fig. 28). In formerly discussed primary diseases a calcification of the vascular walls could not be proved. In only one end-stage kidney did a larger arterial vessel show conspicuous segmental calcium deposits (Fig. 29) as had been reported by Kuzela et al. (1977).

In the literature it is often emphasized that the tubular basement membrane would have a certain predisposition to calcification at the stage of terminal renal insufficiency (Romen 1970; Kuzela et al. 1977; Heptinstall 1983). In contrast to my findings these authors saw metastatic calcifications of the glomerular basement membrane and Bowman's capsule. Calcium phosphate deposits in the glomerular basement membrane were described in a dialysis patient suffering from Wegener's granulomatosis (Sanfilippo et al. 1981).

Regarding the tubular basement membrane my observations agree with references concerning the selective affect on the proximal convoluted tubules on the one hand and on the other a certain simultaneous tubular atrophy. The extraordinary case of Wegener's disease was not accompanied by a considerable calcification of the tubular basement membrane. My observation that the NC is more pronounced in the cortex than in the medulla is in accordance with the literature. Nevertheless no uniform pattern of NC seems to be found in terminal renal insufficiency.

The detachment of the basement membrane and the neoformation of basement membranelike structures described by Romen (1970) is similar to my findings in primary hyperparathyroidism (cf. Fig. 9).

In the literature, interstitial calcifications in end-stage kidneys are an exception to the rule. and are correctly interpreted as a condition resulting from tubular destruction (Kuzela et al. 1977).

In 1985 Goligorsky and coworkers considered treatment with calcium antagonists in cases of uremic NC to reduce the mitochondrial damage caused by calcification. It is doubtful whether a favorable effect can be achieved in already uremic patients.

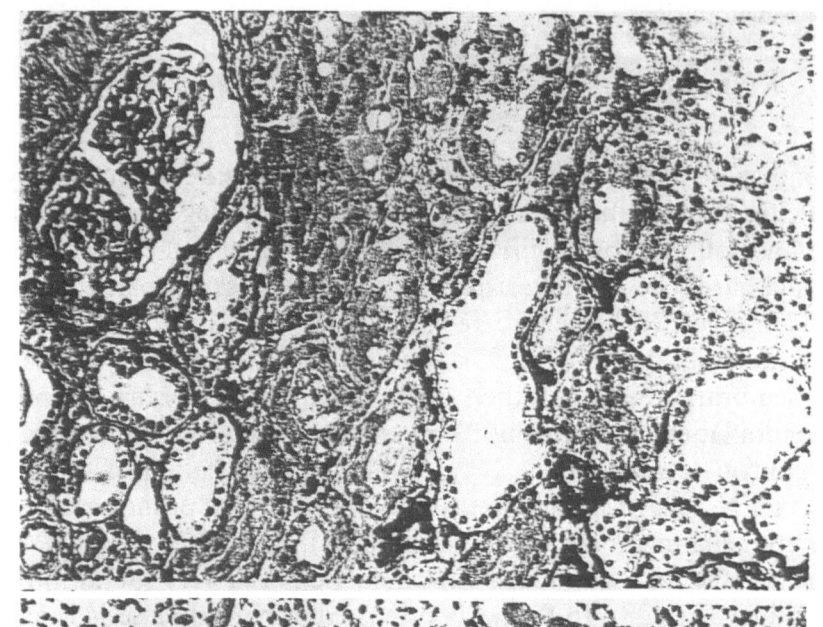

a

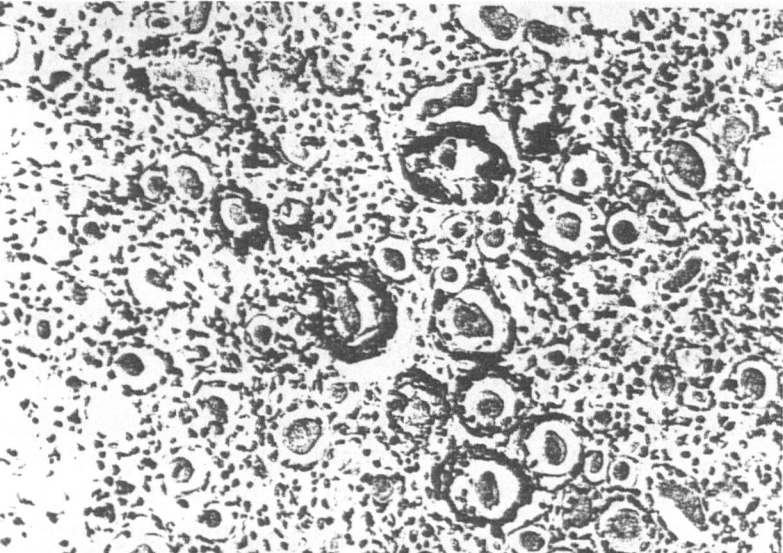

b

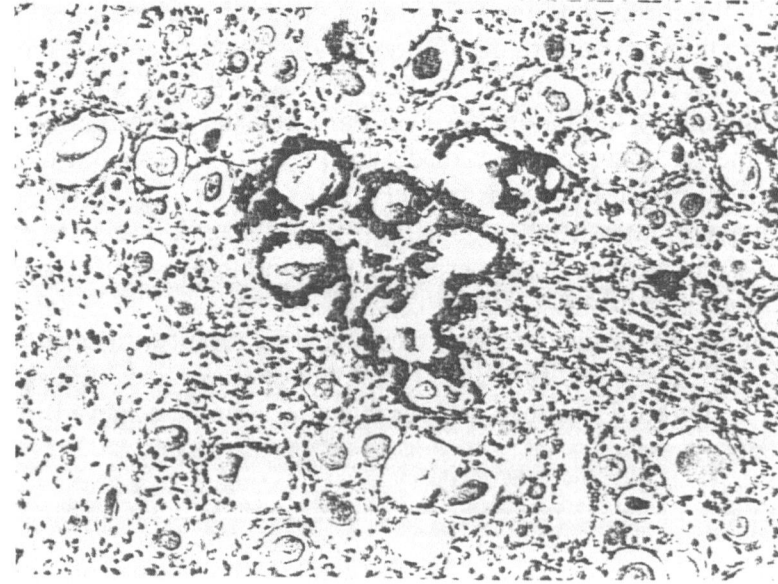

c

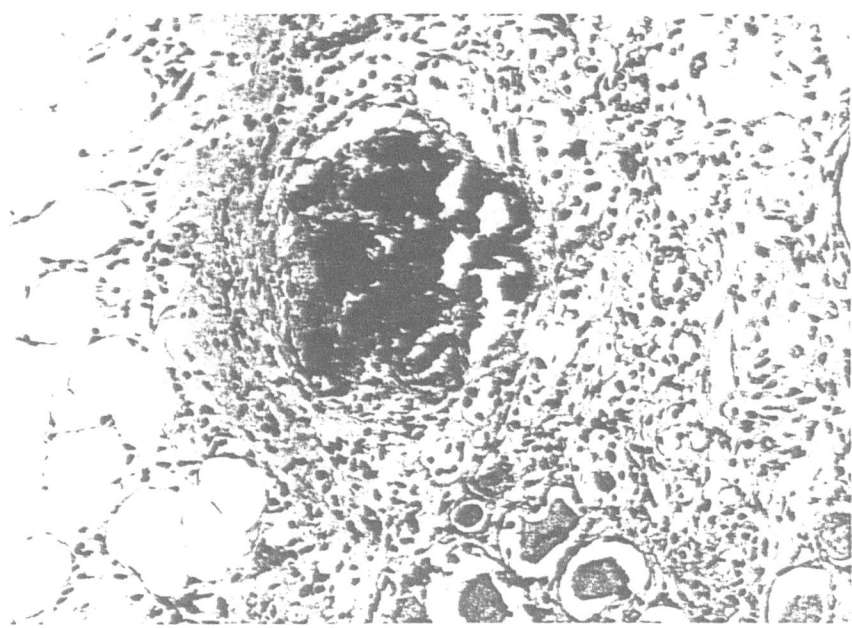

Fig. 28. End-stage kidney: adjacent to a scarred broadened capsula fibrosa is an obliterated almost completely calcified glomerulus. Tubules partly destroyed, partly atrophic. Kossa reaction, × 160

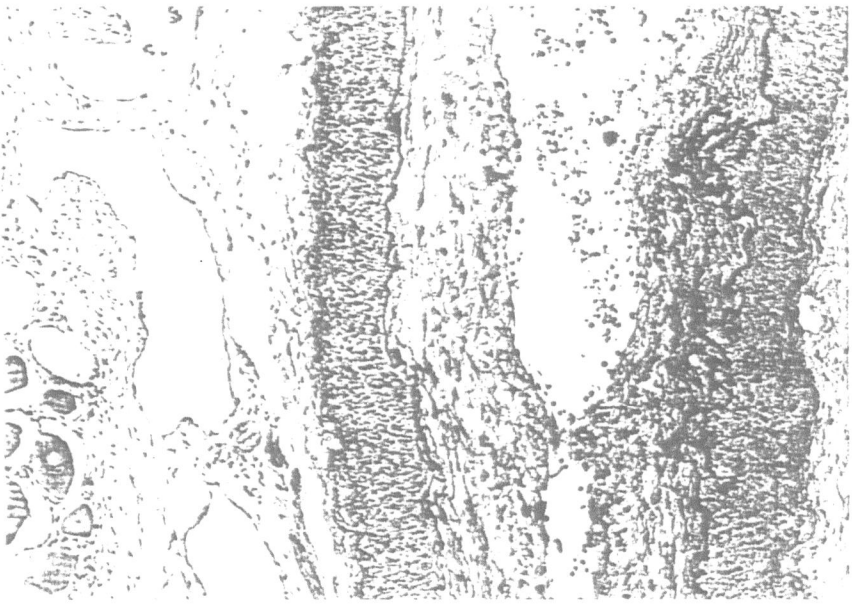

Fig. 29. End-stage kidney: larger renal arterial vessel with segmental elastica calcification. Kossa reaction, × 100

◄ ───

Fig. 27 a–c. End-stage kidney. **a** Renal cortex with slight to moderate, finely granular calcium deposits along the basement membrane of partly atrophic tubules. **b** Another case with pronounced bandlike calcification of the basement membrane, extending to the tubular epithelial cells. **c** In an adjacent area completely calcified epithelial cells of single proximal convoluted tubules. All figures: Kossa reaction, × 160

Autopsy Studies

A detailed light and electron microscopic examination of 100 autopsy cases (Haggitt and Pitcock 1971) demonstrated calcification mainly in the renal medulla. These authors classified slight (37/100), moderate (48/100) and severe (15/100) calcification. In 23% of the cases calcium deposits in the papillae were macroscopically visible. Electron microscopic evaluation of seven cases showed calcium deposits in the interstitium and adjacent to the basement membrane of the collecting ducts, while calcification of mitochondria or other cytoplasmic components was not observed.

Similar results are reported by Herbay and Saeger (1985): of the 980 analyzed autopsy cases, NC was demonstrated – mainly in the papillae – in 77.6% (grade I, 39.4%; grade II, 26.8%; grade III, 8.5%; grade IV, 2.6%).

Different numerical data regarding the incidence of NC might be due mainly to methodological differences (number of sections, methods of calcium detection). The percentages reported by Ungar (1950) for 250 autopsy cases are much lower; NC was proved only in 18.8% of all cases, Randall's plaques type I (intertubularly between collecting ducts) in 9.2%, type II (casts or microliths in lumina of collecting ducts) in about 10%, and simultaneous occurrence in about 3.5%.

The observations are almost consistent in that "normal" autopsy kidneys show smaller calcifications in the medulla and papilla, while calcification in the cortex is both rare and only slightly developed (Cooke 1970).

Topography of Nephrocalcinosis

Comparing my findings with the results reported by Jaccottet (1959) and other authors (cf. Tables 3 and 4) only partial consent exists concerning the different primary diseases. The topographic distribution of calcification is described by Cottier (1980) as heterogeneous; this is evident especially for dystrophic calcification but not for the metastatic type. The emphasis in the literature on the corticomedullary junction or the outer medullary region as places for calcification (Brod 1973; Heidbreder et al. 1974; Cottier 1980; Bichler et al. 1985) cannot as a rule be confirmed in human nephrectomy material. However, this distribution of calcium might be found in many animal experiments (Jaccottet 1959; Shimura 1972; Heidbreder et al. 1974; Shimura et al. 1974; Hennis et al. 1982; Bichler et al. 1985; Hertle et al. 1985). Possible reasons will be discussed in the pathogenesis section.

Pathogenesis of Nephrocalcinosis

While dystrophic NC, i.e., secondary calcification of pathologically altered renal structures or pathologically deposited substances, raises no difficulties in comprehension supposing that serum calcium will be deposited when the pH value is in a neutral or slightly alkaline range (Brod 1973). the circumstances in metastatic NC, however, seem to be more complex. Hypercalcemia and/or hypercalciuria result in an increased calcium load on the renal parenchyma.

The development of NC is certainly determined by several factors; besides the pH value we have to consider the calcium phosphate ratio, the excretion of mucoprotein, the excretion of citrate, and a reduced calcium solubility (Boeckle and Sarre 1976; Heptinstall 1983). Gradation of the renal calcium from the cortex via the medulla to the papilla also seems to be important (Gains et al. 1968; Cooke 1970; Cook and Rosenzweig 1971; Wright and Hodgkinson 1972; Anderson 1982; Schubert 1984). The previous description of predisposed nephron sections, partly in the area of the cortex, however, shows that this calcium gradation is by no means always the determinative factor for the topography of NC.

In accordance with Massry (1979) I believe that some of the pathogenetic mechanisms of NC must remain unexplained. They vary from patient to patient, dependent on the primary conditions. This author pointed out that the duration of hypercalcemia and hypercalciuria, the level of parathormone, the reduced urine acidification and local lesions of renal parenchyma may be determinative factors.

Under ischemia in animal experiments NC occurs particularly in the outer medulla (Hertle et al. 1985). This already mentioned topography, described by several other authors, cannot be explained only by the physiology of renal calcium handling; it must be assumed that factors like the circulatory condition in the medulla, the primary urine flow, the pH value, and other parameters are also involved. Visualizing the different components of the renal parenchyma, the question arises which structures (tubules with basement membrane, epithelial cells and lumina, interstitium, glomeruli, vessel walls) are the most affected by primary and secondary NC.

Experimental observations in the early phase of calcification are essential for the understanding of NC. When calcium is taken up in the tubular epithelial cells, it is accompanied by swelling and disorganization of mitochondria followed by cell decay (Anderson 1982). Toxins induce NC mainly in the proximal tubules, while hypercalcemia leads to calcification in the medulla with focal accentuation of the ascending loops of Henle, the distal convoluted tubules, and, partially, the collecting ducts. Only after severe and long-lasting hypercalcemia the proximal convoluted tubules, glomeruli, vascular walls and interstitium are also affected. Interstitial calcification is attributed partly to a destruction of renal tubules, partly to calcification of basement membranes (compare Cottier 1980).

In an electron microscopic investigation on animals (Caulfield and Schrag 1964), metastatic NC was induced partly by parathormone and partly by calcium gluconate, resulting in a heterogeneous distribution of calcium deposits. The administration of parathormone lead to coarse-granular calcium deposits in the cytoplasm of epithelial cells of distal segments of the proximal tubules, namely in mitochondria and partly in vacuoles (compare Zollinger and Mihatsch 1978). Parathormone is thought to stimulate the mitochondria, which are structurally altered by calcium overloading. The active role of mitochondria has been emphasized by other authors as well (Gerlach and Themann 1965; Schäfer 1979); this opinion, however, has been disputed. Other authors regard prior functional and structural damage of the mitochondria as important conditions for calcification (Scarpelli 1965; Zollinger and Mihatsch 1978). In vitro findings (Mergner et al. 1977), however, indicate an energy-consuming calcium uptake by the mitochondria and justify the terms primary and metastatic NC.

It is noteworthy that many tubular epithelial cells can be totally free of calcium. Usually intracellular calcium deposits precede intraluminal ones, but the reverse relation has been reported as well (Cottier 1980).

In some cases, severely mineralized mitochondria and vacuoles merge together and large amounts of calcium can be released with apical cytoplasmic portions into the tubular lumen (Jaccottet 1959; Caulfield and Schrag 1964), hence luminal cell borders are difficult to define exactly. The fact that the nucleus itself is supposed not to be calcified might explain the regenerative power of damaged tubular epithelial cells. The material delivered into the tubular lumen is mainly found in the distal tubules; however, whole tubular epithelial cells may also be discharged into the tubular lumen (Duffy et al. 1971).

Reversibility of Nephrocalcinosis

The question of NC reversibility is interesting in cases where normalization of the calcium handling at least theoretically might lead to the disappearance of the calcification in renal tissue. With regard to the application of calcium antagonists it has to be assumed that they will hardly be able to reverse an already present calcification of several renal structures.

The question of the possibility of reversing NC cannot be answered definitively (Brod 1973), although positive results have been obtained in animal experiments (Pyrah 1949; Sanderson 1959; Carone et al. 1960; Caulfield and Schrag 1964; Khan et al. 1982). In corresponding experiments, NC has been induced by different methods; the regeneration process after cessation of calcium overload turned out to be more favorable for less distinct NC. Reversibility actually means healing by regeneration of the epithelial cells. If intratubular casts have already developed, the alterations must be regarded as irreversible (Carone et al. 1960).

Only metastatic NC is relevant; the disappearance of dystrophic calcification from irreversibly damaged renal structures has no clinical significance, except as a possible pacemaker function for kidney stone formation.

Relationship Between Nephrocalcinosis and Nephrolithiasis

It could be suggested that a relation exists between renal parenchyma calcification and calcium containing concrements in the pelvicalyceal system. In different primary diseases NC and nephrolithiasis are seen simultaneously; in primary hyperparathyroidism the figures for NC and/or nephrolithiasis range from 60 to 70% (Vahlensieck 1979; Cottier 1980; Haas and Dambacher 1980; Heptinstall 1983). The development of nephrolithiasis is more probable the longer the duration of NC (Pyrah 1949; Heptinstall 1983).

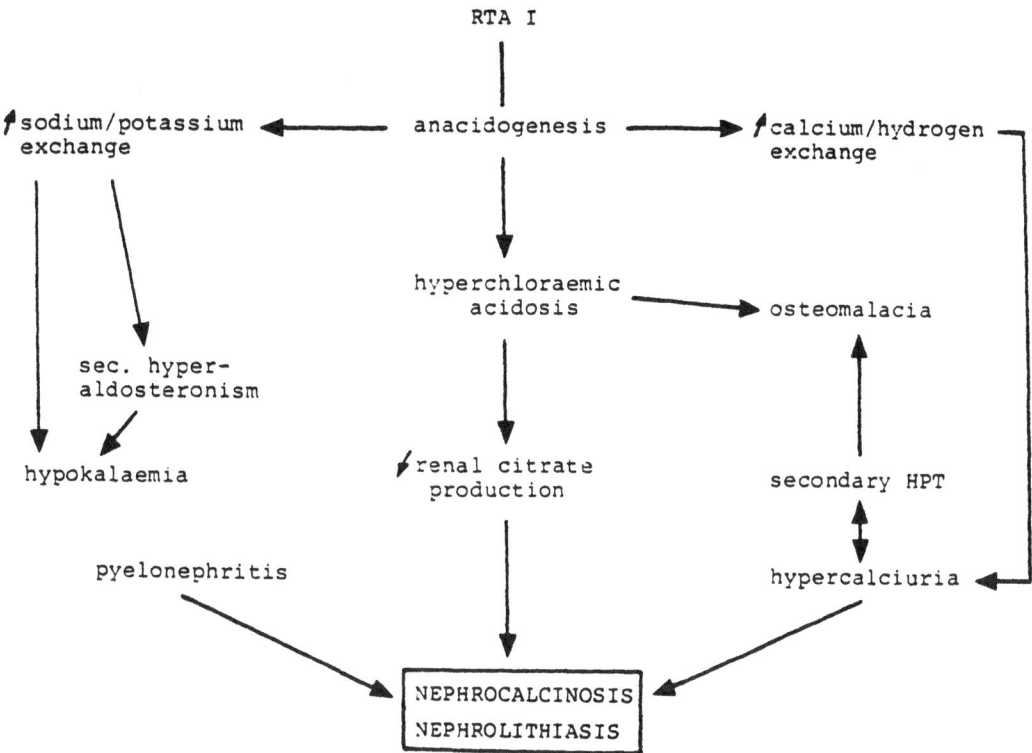

Fig. 30. Pathogenesis of nephrocalcinosis and nephrolithiasis in renal tubular acidosis (type I). (Modified after Gerok W, *Primäre Tubulopathien*; Thieme, Stuttgart 1969)

The simultaneous or successive occurrence of NC and of nephrolithiasis has also been described for renal tubular acidosis (Brod 1973; Marquardt 1973; Gerok 1976; Sommerkamp 1980). Of all cases of renal tubular acidosis, 73% are accompanied by NC and nephrolithiasis (Marquardt 1973). According to Gerok (1976), who presented the essential pathogenetic mechanisms schematically (Fig. 30), pyelonephritis increases the risk of nephrolithiasis.

More significant in the difficult discussion of the causality, however, are the numerous cases where a distinct NC is not accompanied by nephrolithiasis or nephrolithiasis not by NC.

Although these two alterations are often found together, Anderson (1982) pointed out that he did not regard them as extreme variants of the same kind of condition and that NC does not necessarily induce nephrolithiasis. Of particular relevance in metastatic NC are alterations of calcium metabolism in the renal parenchyma at the cellular level, whereas in the development of nephrolithiasis, precipitation of factors in the supersatured urine are thought to be significant.

Comparing the etiology of nephro(uro)lithiasis (Table 5) with that of NC (Table 2), we find a conspicuous correspondence. It must be mentioned, however, that additional factors may also be significant in the etiopathology of nephrolithiasis, for example epidemiologic factors like genetic heredity, age, sex, geographic conditions, alimentation, and water consumption (general review: Drach 1986).

With regard to the etiological and pathogenetic factors of nephrolithiasis, e.g., infections or urodynamic causes like urinary tract obstruction, reflux or urinary

Table 5. Etiology of nephro(uro)lithiasis

A. *Epidemiological factors*

B. *Primary nonurological diseases*
   - Hyperparathyroidism
   - Sarcoidosis (Boeck's disease)
   - Primary or secondary hyperoxaluria
   - Hypercortisolism
     - Cushing's disease
     - Systemic cortisone therapy
   - Vitamin D intoxication
   - Malignant bone tumors $\pm$ skeletal metastases
   - Myeloproliferative diseases
   - Milk alkali syndrome
   - Immobilization
   - Cytostatic therapy, radiotherapy
   - Colitis ulcerosa
   - Crohn's disease
   - Status after small intestine resection
   - Dehydration
   - Abuse of laxatives
   - Gout
   - Vitamin A deficiency

C. *Altered renal function and lesions of the urinary tract*
   - Renal tubular acidosis
   - Urodynamic causes
     - Pelvi-ureteral obstruction
     - Distal ureteral obstruction
     - Vesico-ureteral reflux
     - Infravesical obstruction
     - Double ureter
     - Ureterocele
     - Aberrant vessel
     - Combination of the mentioned causes
   - Renal infections
   - Medullary sponge kidney
   - Neurogeneous bladder
   - Cystic kidneys
   - Cystinuria

tract duplications, some papers should be referred to for further information (Brod 1973; Coe and Kavalach 1974; Marquardt and Nagel 1977; Smith 1978; Reiner et al. 1979; Sinno et al. 1979; Drach 1986). In discussing single factors it should be considered that nephrolithiasis is influenced by many factors (Hautmann and Lutzeyer 1980) and that a number of the above-mentioned factors are of no importance for NC.

Several theories on lithogenesis have been discussed from the pathogenetic point of view. They concern crystallisation, crystal aggregation, fixed and free formation of particles, supersaturation of the urine, pH value of the urine, inhibitors, stone matrix, uromucoid, and vitamin A deficiency (basic literature: Bichler et al. 1976; Altwein et al. 1977; Finlayson 1978; Bommer et al. 1979; Vahlensieck

1979; Cottier 1980; Hautmann and Lutzeyer 1980; Vahlensieck et al. 1982; Bichler et al. 1983; Coe and Parks 1984; Smith 1984; Drach 1986).

Randall's theory (Randall and Melvin 1937) has been judged differently in the literature. The first description of Randall's plaques distinguished between interstitial calcifications close to the papillary surface (type I) and an intratubular formation of microliths (type II). Close interrelations between this type of manifestation of NC with nephrolithiasis have been partly questioned (Anderson and McDonald 1946; Cooke 1970; Hautmann and Lutzeyer 1980), and partly affirmed (Gains et al. 1968; Vermeulen and Lyon 1968; Anderson 1982; Schubert 1984; Smith 1984; Drach 1986).

As regards the possibility of lithogenesis, the urodynamic situation seems to be more important and is thought to be impaired in 60 to 70% of nephrolithiasis cases (Vahlensieck 1979). The fast transport of urine from the renal pelvis via the ureter into the bladder does not permit stone formation in the renal pelvis up to a diameter which would prevent the passage of a small concrement from the renal pelvis (Hautmann and Lutzeyer 1980; Drach 1986). From this point of view the mechanism of adherence and enlargement of particles at the papillary surface or in the fornical region seems to be of special relevance.

Observations of cases where the development of nephrolithiasis would be expected but does not actually occur are of interest. In a case studied by me of massive idiopathic osteolysis with hypercalciuria lasting 10 years and with distinct NC, no concrement developed. The patient's normal kidney function until death might have prevented it.

Summary and Conclusions

Using a comprehensive definition each calcium deposit in the renal parenchyma has to be called NC. A primary or metastatic type is distinguished from a secondary or dystrophic one, supposing that a primary NC can develop into a secondary one at an early stage. The whole spectrum of etiological factors is presented and documented by light and electron microscopic findings and is compared critically with the literature. It is difficult to give a sufficiently general statement about the predisposition of nephron segments or other renal structures in primary diseases. This indicates a complex process determined by many factors. Primary and secondary NC mainly affect the tubular system, while interstitial, glomerular, and vascular calcifications are not of importance.

The application of calcium antagonists seems to be useful to prevent further calcium deposition in certain primary diseases. A spontaneous reversibility of metastatic calcium deposits can be only expected in the early phase of NC. The discussion of the pathogenesis of NC requires consideration of the possible relationship with nephrolithiasis.

References

Altwein JE. Alken P, Joost J (1977) Probleme der Nierensteinerkrankung. Z Allg Med 53:1079–1090

Anderson CK (1982) Pathology of nephrocalcinosis and stone formation. In: Chisholm GD, Williams DI (eds) Scientific foundations of urology, 2nd edn. William Heinemann Medical Books Ltd., London, pp 296ff.

Anderson L, McDonald JR (1946) The origin, frequency, and significance of microscopic calculi in the kidney. Surg Gynec Obstet 82:275–282

Bichler KH, Kirchner C, Ideler V (1976) Uromucoid excretion of normal individuals and stone formers. Br J Urol 47:733–738

Bichler KH. Kirchner C, Weiser H, Korn S, Strohmaier W, Schmitz-Moormann P, Hanck A, Nelde HJ (1983) Influence of vitamin A deficiency on the excretion of uromucoid and other substances in the urine of rats. Clin Nephrol 20:32–39

Bichler KH. Strohmaier WL, Schanz F, Nelde HJ, Gaiser I, Schulze E, Schreiber M (1985) Zur Wirkung von Kalziumantagonisten (Nifedipin) auf die Nephrokalzinose und Kalziumausscheidung der Ratte. Urol Int 40:13–21

Boeckle H, Sarre H (1976) Nephrokalzinose. In: Sarre H (Hrsg) Nierenkrankheiten. Physiologie, Pathophysiologie, Untersuchungsmethoden. Klinik und Therapie. Thieme. Stuttgart, S 235ff.

Bommer J, Ritz E, Tschöpe W, Waldherr R. Gebhardt M (1979) Urinary matrix calculi consisting of microfibrillar protein in patients on maintenance hemodialysis. Kidney Int 16:722–728

Brod J (1973) The kidney. Butterworth & Co. Ltd.. London, pp 580f.. 585ff.. 588ff.

Bulger RE, Dobyan DC (1984) Pathology of acute renal failure. In: Robinson RR (ed) Nephrology, vol I. Proceedings of the IXth International Congress of Nephrology. Springer, Berlin Heidelberg New York Tokyo, pp 715ff.

Bunting H (1951) Histochemical analysis of pathological mineral deposits at various sites. Arch Pathol 52:458–469

Carone FA. Epstein FH, Beck D, Levitin H (1960) The effects upon the kidney transient hypercalcemia induced by parathyroid extract. Am J Pathol 36:77–89

Caulfield JB, Schrag PE (1964) Electron microscopic study of renal calcification. Am J Pathol 44:365–381

Coe FL. Kavalach AG (1974) Hypercalciuria and hyperuricosuria in patients with calcium nephrolithiasis. N Eng J Med 291:1344–1350

Coe FL. Parks JH (1984) Urolithiasis. Pathogenesis of calcium renal stones. In: Robinson RR (ed) Nephrology, vol II. Proceedings of the IXth International Congress of Nephrology. Springer. Berlin Heidelberg New York Tokyo. pp 980ff.

Cooke SAR (1970) The site of calcification in the human renal papilla. Br J Surg 57:890–896

Cooke SAR, Rosenzweig D (1971) The concentration of calcium in the human renal papilla and the tendency to form calcium containing renal stones. Nephron 8:528–539

Cottier H (1980) Pathogenese, Bd 1. Springer. Berlin Heidelberg New York, S 793ff., 817ff., 824ff.

Drach GW (1986) Urinary lithiasis. In: Walsh PC, Gittes RF, Perlmutter AD, Stamey TA (eds) Campbell's urology, vol 1, chap 25, 5th edn. Saunders, Philadelphia London Toronto Mexico City Rio de Janeiro Sydney Tokyo Hong Kong, pp 1093ff.

Duffy JL, Suzuki Y, Churg J (1971) Acute calcium nephropathy early proximal tubular changes in the rat kidney. Arch Pathol 91:340–350

Engfeldt B, Rhodin J, Strandh J (1962) Studies of the kidney ultrastructure in hypervitaminosis D. Acta Chir Scand 123:145–147

Epstein FH (1971) Calcium nephropathy. In: Strauss MB. Welt LG (eds) Diseases of the kidney. Little, Brown and Co., Boston, pp 903ff.

Finlayson B (1978) Physicochemical aspects of urolithiasis. Kidney Int 13:344–360

Gains NA. Michaels CW, Thwaites MZ, Trounce JR (1968) Calcium concentration in the kidney. Nephron 5:352–361

Gerlach U, Themann H (1965) Elektronenmikroskopische Untersuchung der metastatischen Calcifizierung. Klin Wochenschr 43:1262–1268

Gerok W (1976) Primäre Störungen tubulärer Partialfunktionen (Primäre Tubulopathien). In: Sarre H (Hrsg) Nierenkrankheiten. Physiologie, Pathophysiologie, Untersuchungsmethoden, Klinik und Therapie. Thieme, Stuttgart, S 363ff.

Goligorsky MS, Chaimovitz C, Rapoport J, Goldstein J, Kol R (1985) Calcium metabolism in uremic nephrocalcinosis: preventive effect of verapamil. Kidney Int 27:774–779

Haas HG, Dambacher MA (1980) Hyperparathyreoidismus und Harnsteinbildung. In: Vahlensieck W (Hrsg) Urolithiasis 4. Der Kalzium-Phosphat-Stein. Springer, Berlin Heidelberg New York, S 48ff.

Haas HG, Dambacher MA (1982) Calciumhormone, Skelett und Mineralstoffwechsel. In: Siegenthaler W (Hrsg) Klinische Pathophysiologie. Thieme, Stuttgart New York, S 349ff.

Haggitt RC, Pitcock JA (1971) Renal medullary calcifications: a light and electron microscopic study. J Urol 106:342–347

Hautmann R, Lutzeyer W (1980) Spezielle Pathogenese des Kalziumoxalatsteins. In: Vahlensieck W (Hrsg) Urolithiasis 3. Der Kalzium-Oxalat-Stein. Springer, Berlin Heidelberg New York, pp 1ff.

Heidbreder E, Hennemann H, Heidland A (1974) Hypercalciurie – Nephrocalcinose – Kalklithiasis der Niere. Dtsch Med Wochenschr 99:586–590

Hennis HL, Hennigar GR, Greene WB, Hilton CW, Spector M (1982) Intratubular calcium phosphate deposition in acute analgesic nephropathy in rabbits. Am J Pathol 106:356–363

Heptinstall RH (1983) Pathology of the kidney, vol 2, 3rd edn. Little, Brown and Co., Boston Toronto, pp 623f.

Heptinstall RH (1983) Pathology of the kidney, vol 3, 3rd edn. Little, Brown and Co., Boston Toronto, pp 1599ff.

Herbay A, Saeger W (1985) Die Nephrocalcinose im unausgewählten Sektionsgut. Ber Path 101:285

Hertle L, Garthoff B, Chur C, Pötz B, Funke PJ, Senge T (1985) Limitierung des ischämischen Nierenschadens durch den Calciumantagonisten Nisoldipin. In: Harzmann R, Jacobi GH, Weißbach L (Hrsg) Experimentelle Urologie. Springer, Berlin Heidelberg New York Tokyo, S 189ff.

Jaccottet MA (1959) Zur Histologie und Pathogenese der Nierenverkalkung (Nephrocalcinose und dystrophische Kalknephrose). Virchows Arch Path Anat 332:245–263

Khan SR, Finlayson B, Hackett RL (1982) Experimental calcium oxalate nephrolithiasis in the rat. Role of the renal papilla. Am J Pathol 107:59–69

Klatskin G, Gordon M (1953) Renal complications of sarcoidosis and their relationship to hypercalcemia. With a report of two cases simulating hyperparathyroidism. Am J Med 15:484–498

Kossa von J (1901) Über die im Organismus künstlich erzeugbaren Verkalkungen. Beitr Path Anat 29:163–202

Kuzela DC, Huffer WE, Conger JD, Winter SD, Hammond WS (1977) Soft tissue calcification in chronic dialysis patients. Am J Pathol 86:403–424

Laberke HG (1987) Die Refluxnephropathie. Ätiologie, Pathogenese und Differentialdiagnose. Veröffentlichungen aus der Pathologie, Bd 128. Fischer, Stuttgart

Magori A, Ormos J, Fazekas M, Siklos L, Sonkodi S, Rudas L, Turi S (1983) Concretions in renal basement membranes. Diagn Histopath 6:195–202

Marquardt H (1973) Inkomplette renale tubuläre Acidose bei rezidivierender Nephrolithiasis und Nephrocalcinose. Urologe A 12:162–166

Marquardt H, Nagel R (1977) Urolithiasis in childhood. Urology 9:627–629

Massry SG (1979) Effects of electrolyte disorders on the kidney. In: Earley LE, Gottschalk CW (eds) Strauss und Welt's diseases of the kidney, vol 2, 3rd edn. Little, Brown and Co., Boston, pp 1403ff.

Mergner WJ, Smith MW, Sahaphong S, Trump BF (1977) Studies on the pathogenesis of ischemic cell injury. VI. Accumulation of calcium by isolated mitochondria of ischemic rat kidney cortex. Virchows Arch B 26:1–16

Pyrah LN (1949) Tubular calcification in the kidney. Br J Urol 21:27–29

Randall A, Melvin PD (1937) The morphology of renal calculus. J Urol 37:737–745

Reiner RJ, Kroovand RL, Perlmutter AD (1979) Unusual aspects of urinary calculi in children. J Urol 121:480–481

Reubi F (1970) Tubulare und interstitielle Erkrankungen. Die Nephrokalzinose. In: Nieren-krankheiten. 2. Aufl. Huber, Bern Stuttgart Wien, S 370ff.

Romen W (1970) Basalmembranverkalkungen in der Niere. Pathomorphologische Beobachtungen an menschlichen Schrumpfnieren. Virchows Arch A 350:240–249

Ross L, Chin W (1970) Metastatic calcification of renal glomerular basement-membrane. J Pathol 101:69–71

Sanderson PH (1959) Functional aspects of renal calcification in rats. Clin Sci 18:67–79

Sanfilippo F, Wisseman C, Ingram P, Shelburne J (1981) Crystalline deposits of calcium and phosphorus. Their appearance in glomerular basement membranes in a patient with renal failure. Arch Pathol Lab Med 105:594–598

Scarpelli DG (1965) Experimental nephrocalcinosis. A biochemical and morphologic study. Lab Invest 14:123–141

Schäfer HJ (1979) Zellcalcium und Zellfunktion. Veröffentlichungen aus der Pathologie. Progress in Pathology. Fischer. Stuttgart New York, S 30ff., 147ff.

Scholz DA, Keating FR Jr (1956) Renal insufficiency, renal calculi and nephrocalcinosis in sarcoidosis. Am J Med 21:75–84

Schubert GE (1984) Niere und ableitende Harnwege. In: Remmele W (Hrsg) Pathologie, Bd 3. Springer, Berlin Heidelberg New York Tokyo, S 1ff.

Seifert G, Schäfer HJ, Schulz A (1975) Die Bedeutung des intrazellulären Calciumtransportes für die Zellfunktion. Dtsch Med Wochenschr 100:1854–1862

Shimamura T (1972) Drug-induced renal medullary necrosis. I. Structural alteration of renal medulla. Arch Path 94:406–410

Shimamura T, Aogaichi M, Liu CY (1974) Drug-induced renal medullary necrosis. II. Mode of calcification in the kidney. Exp Molec Pathol 20:109–114

Sinno K, Boyce WH, Resnick MI (1979) Childhood urolithiasis. J Urol 121:662–664

Smith LH (1978) Calcium-containing renal stones. Kidney Int 13:383–389

Smith LH (1984) Physicochemical factors in calcium oxalate urolithiasis. In: Robinson RR (ed) Nephrology, vol II. Proceedings of the IXth International Congress of Nephrology. Springer, Berlin Heidelberg New York Tokyo, pp 990ff.

Sommerkamp H (1980) Steinbildung bei renal-tubulärer Azidose. In: Vahlensieck W (Hrsg) Urolithiasis 4. Der Kalzium-Phosphat-Stein. Springer, Berlin Heidelberg New York, S 59ff.

Ungar H (1950) Calcium deposits in renal papillae. Arch Path 49:687–698

Vahlensieck W (1979) Epidemiologie. Allgemeine Kausal- und Formalgenese. Diagnostik. In: Vahlensieck W (Hrsg) Urolithiasis 1. Springer. Berlin Heidelberg New York. S 1ff.

Vahlensieck W, Bach D, Hesse A, Strenge A (1982) Epidemiology, pathogenesis and diagnosis of calcium oxalate urolithiasis. Int Urol Nephrol 14:333–347

Vermeulen CW, Lyon ES (1968) Mechanisms of genesis and growth of calculi. Am J Med 45:684–692

Virchow R (1855) Kalkmetastasen. Virchows Arch Path Anat 8:103–113

Weller RO, Nester B, Cooke SAR (1972) Calcification in the human renal papilla: an electron-microscope study. J Pathol 107:211–216

Wright RJ, Hodgkinson A (1972) Oxalic acid, calcium, and phosphorus in the renal papilla of normal and stone-forming rats. Invest Urol 9:369–375

Zollinger HU, Mihatsch MJ (1978) Renal pathology in biopsy. Light, electron and immuno-fluorescent microscopy and clinical aspects. Springer, Berlin Heidelberg New York, pp 483ff.

Clinical Aspects of Nephrocalcinosis

R. A. L. Sutton[1]

Introduction

The term nephrocalcinosis was coined by Albright et al. [1] to describe the calcium deposits found in the kidneys in hyperparathyroidism. It is distinct from nephrolithiasis, or stones in the pelvicalyceal system, though the two conditions frequently coexist.

Nephrocalcinosis may be cortical or medullary; it may result from abnormalities in the chemistry of the blood (as in hypercalcemia) or in the urine (as in renal tubular acidosis). Nephrocalcinosis may also be "dystrophic" in nature, as in cortical and papillary necrosis, or related to anatomical abnormalities in the kidney, as in medullary sponge kidney (tubular ectasia). Nephrocalcinosis may be evident in simple radiographs, or may require more sensitive methods such as ultrasound [2], contact radiography, or histology for its detection.

The pathology and histology of nephrocalcinosis has been examined by Anderson [3]. Nephrocalcinosis appears to result from varying combinations of intraluminal precipitation, deposition within tubular epithelial cells, and interstitial, peritubular deposition of calcium salts. Focal, interstitial, calcified lesions immediately beneath the surface epithelium of the pyramids were described by Randall [4] and have been proposed as foci upon which renal calculi may form.

Causes of Nephrocalcinosis

Wrong [5, 6] has summarized his extensive experience of nephrocalcinosis. Table 1 shows the diagnosis in his 194 cases of radiological nephrocalcinosis. In only 6.5% of cases was the cause not discovered.

[1] The University of British Columbia, Faculty of Medicine, Department of Medicine, 910 West, 10th Avenue, Vancouver, BC, V5Z 1M9, Canada.

Nephrocalcinosis, Calcium Antagonists, and Kidney
Ed. by K.-H. Bichler and W.L. Strohmaier
© Springer-Verlag Berlin Heidelberg 1988

Table 1. Diagnostic categories in 194 patients with radiological nephrocalcinosis. (After Wrong [6])

	%
Cortical	2
Cortical necrosis	
Chronic glomerulonephritis	
Hemolytic-uremic syndrome	
Rejected renal transplants	
Medullary	
Hyperparathyroidism	35
Renal tubular acidosis	24
Complete ($^2/_3$)	
Incomplete ($^1/_3$)	
Tubular ectasia	8
Undiscovered cause	6.5
Milk-alkali syndrome	5.5
Idiopathic hypercalciuria	5
Papillary necrosis	3.5
Sarcoidosis	2.5
Osteoporosis	2.5
Oxalosis	2.5
Chronic pyelonephritis (iatrogenic)	1
Hypervitaminosis D	1
Acetazolamide	0.5
Infantile hypercalcemia	0.5

Specific Types of Nephrocalcinosis

I. Cortical Nephrocalcinosis

This form of nephrocalcinosis is generally dystrophic, resulting from severe disease (or necrosis) of cortical structures. The cortical calcification which follows acute cortical necrosis of pregnancy, especially after prolonged hemodialysis treatment, may be dense, and may have a double or tram-line configuration [7]. Cortical calcification may also occur in chronic glomerulonephritis and in the hemolytic uremic syndrome [6] and may be associated with medullary nephrocalcinosis, for example in hyperoxaluria and oxalosis, and occasionally in primary hyperparathyroidism.

II. Medullary Nephrocalcinosis

By far the commonest causes of medullary nephrocalcinosis are primary hyperparathyroidism and renal tubular acidosis. In both disorders nephrocalcinosis may be associated with nephrolithiasis.

a) Primary Hyperparathyroidism. Nephrocalcinosis has been reported in 16.5% of a large series of 700 patients with primary hyperparathyroidism [6]. Although nephrocalcinosis and nephrolithiasis may coexist, more commonly they occur independently. Patients with nephrocalcinosis often have some reduction in glomerular filtration rate, and a normal or low urinary calcium excretion, and may have associated osteitis fibrosa cystica, whereas patients presenting with renal calculi usually have hypercalciuria and do not have overt hyperparathyroid bone disease. The reasons for these different clinical presentations in primary hyperparathyroidism remain unclear: the suggestion that they may relate to differences in calcitriol production [8] has not been confirmed [9].

b) Renal Tubular Acidosis (RTA). RTA is now subdivided into three major types – RTA-1, RTA-2, and RTA-4. Nephrocalcinosis is probably associated only with RTA-1 [10], or classical distal RTA of the complete or incomplete variety [11]. RTA-1 is more common in females than males, and is usually associated with auto-immune disease, most often Sjogren's disease. The nephrocalcinosis may be extensive and may be associated with recurrent nephrolithiasis (Figs. 1 and 2). Since the metabolic acidosis is present without nephrocalcinosis in RTA-2 and RTA-4, factors other than metabolic acidosis must be responsible for the nephrocalcinosis. Hypercalciuria is in inconsistent feature of RTA-1; in a recent series of 29 patients the mean urinary calcium excretion was only 161 mg [12]. When hypercalciuria is present it may result from bone resorption secondary to metabolic acidosis or from the direct inhibitory effect of metabolic acidosis on renal tubular calcium transport [13] or from resistance to the effect of parathyroid hormone to enhance tubular calcium reabsorption [14]. Bicarbonate delivery to the distal nephron promotes renal calcium reabsorption [13] and this may in part account for the lack of hypercalciuria and nephrocalcinosis in RTA-2. A high

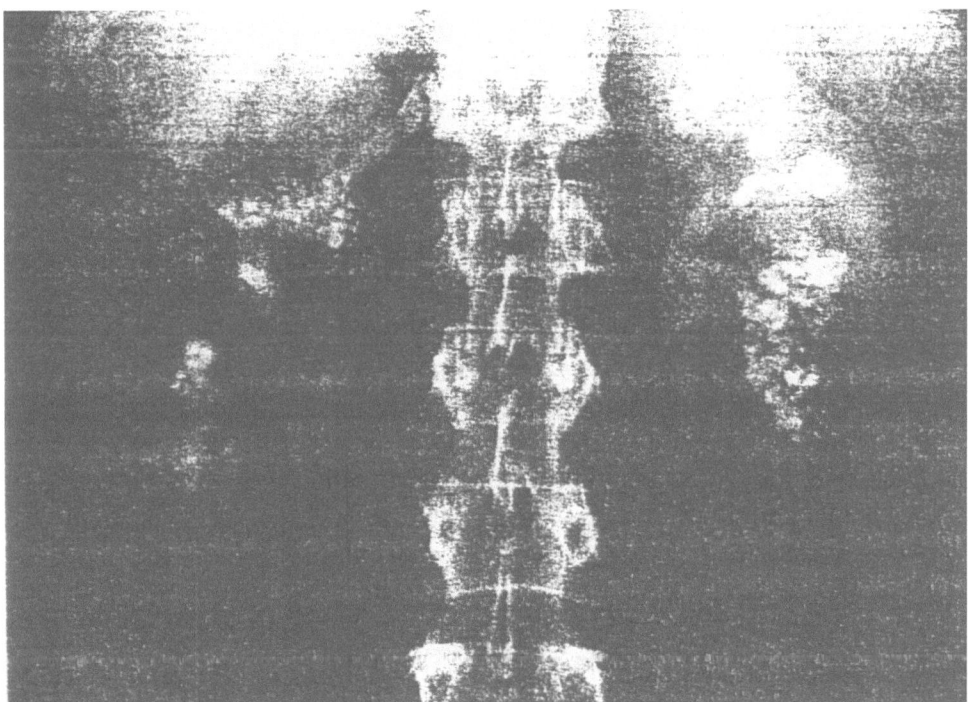

Fig. 1. Excretory urogram in a patient with type I renal tubular acidosis, showing dense bilateral medullary nephrocalcinosis

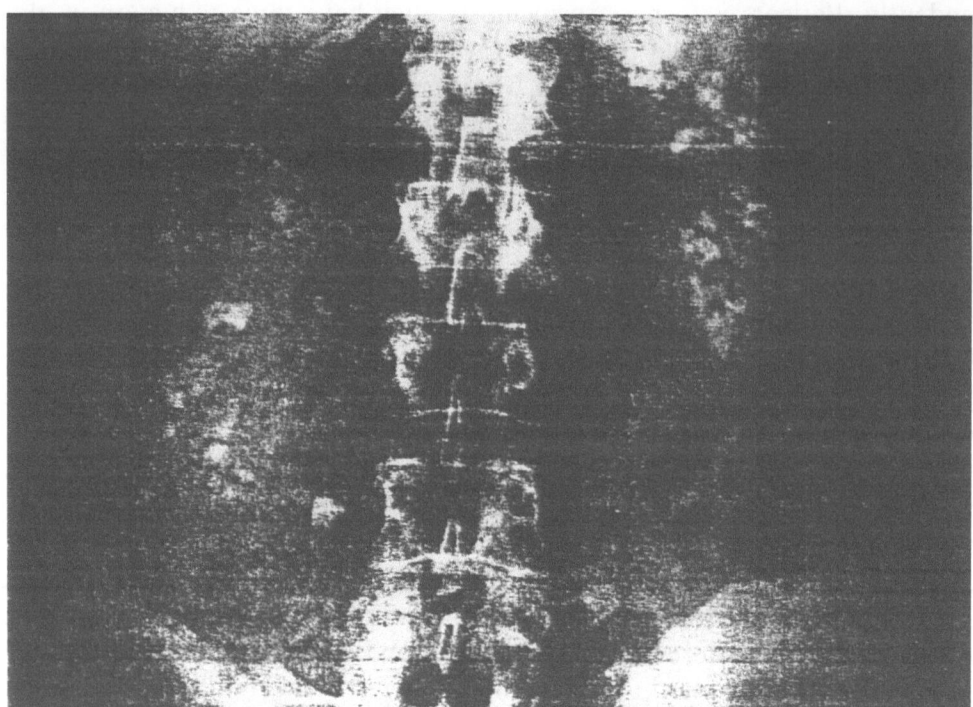

Fig. 2. Plain abdominal film in same patient as Fig. 1. A renal calculus has obstructed the right pelvi-ureteric junction, causing a right hydronephrosis with separation of the medullary calcifications

urine pH is unlikely to account for nephrocalcinosis in RTA–1 since the average urine pH ultimately reflects the generation of hydrogen ions from the diet, and furthermore treatment with alkalis, which presumably raises urine pH, reduced the incidence of stones [15]. The principal cause of nephrolithiasis and renal stones in RTA–1 is probably hypocitraturia [12]. It is suggested that sufficient oral base therapy should be given to ensure a normal urinary citrate excretion. Hypocitraturia has been attributed to intracellular acidosis and to potassium depletion; however, it is also present in incomplete RTA [16] and other factors may therefore be important in causing the reduced citrate excretion.

c) Tubular Ectasia (Medullary Sponge Kidney). This condition differs from other types of medullary nephrocalcinosis in that the calcification, although outside the pelvicalyceal system, is intraluminal, occupying cystic dilatations of the terminal collecting ducts. It may be suspected if X-rays show multiple well-defined round or oval calcifications in the regions of the renal papillae (Fig. 3). It is diagnosed by excretory urography, which shows the dilated collecting ducts (Fig. 4), but not by retrograde urography. It may be unilateral or bilateral, focal or diffuse. The incidence depends on the quality of the excretory urogram; Yendt and colleagues reported its presence in 21% of patients with calcium nephrolithiasis [17], while other groups find a much lower incidence [18]. Tubular ectasia is sometimes associated with hypercalciuria and with RTA–1 [6]; the extent to which these factors contribute to the development of calcifications in the dilated collecting ducts is not certain. The intrapapillary calculi consist mainly of calcium phosphate: inter-

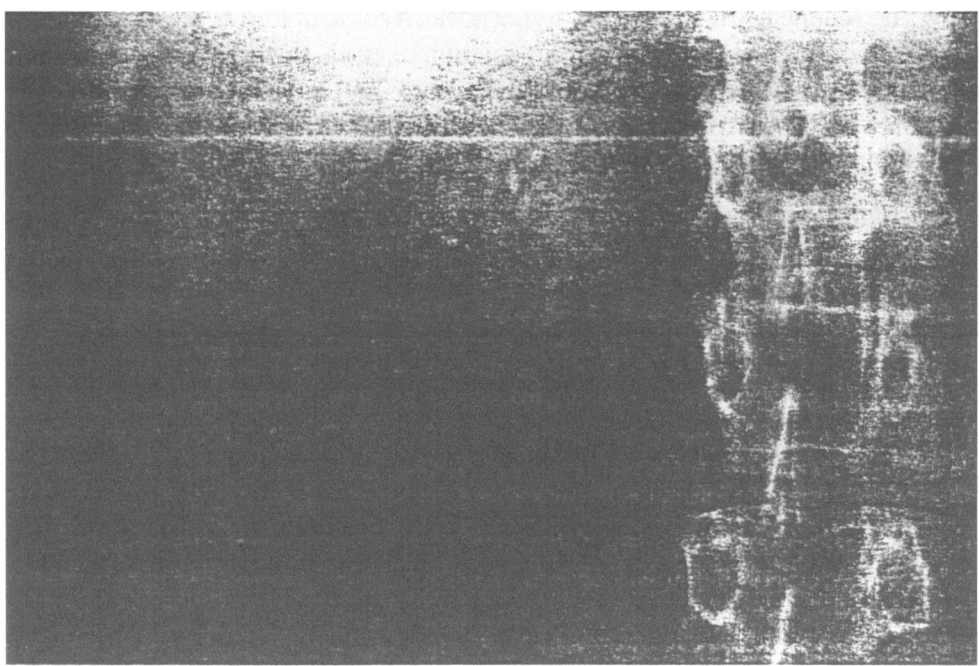

Fig. 3. Plain abdominal film showing medullary nephrocalcinosis in a patient with medullary sponge kidney

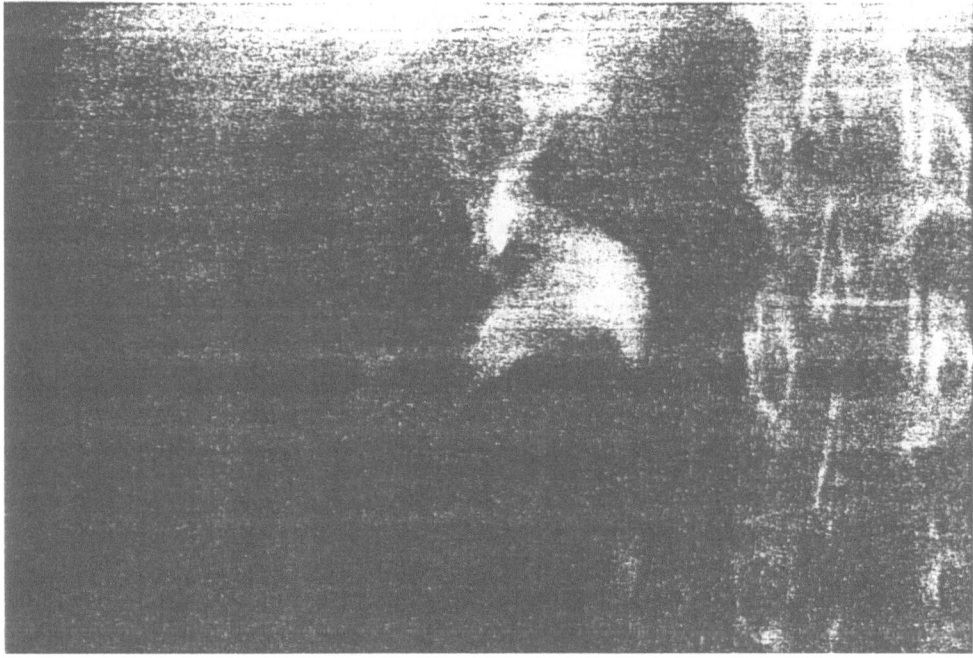

Fig. 4. Excretory urogram in same patient as Fig. 3, showing dilated collecting ducts (tubular ectasia) in the upper papillae at the location of the calcifications

estingly, the renal calculi in primary hyperparathyroidism and RTA–1 also more often consist of calcium phosphate than calcium oxalate, the principal constituent of stones in idiopathic nephrolithiasis. Parks et al. [19] have recently pointed out that tubular ectasia, particularly coupled with urinary tract infection, is a cause of a particularly severe form of renal stone disease, especially in women.

d) Milk-Alkali Syndrome. In this condition, renal impairment is associated with nephrocalcinosis and hypercalcemia and/or hypercalciuria in patients taking excessive quantities of absorbable calcium salts, most often calcium carbonate. With the current tendency to ingest calcium supplements, often calcium carbonate, in the hope of preventing osteoporosis, this condition may be seen more frequently in the future.

e) Idiopathic Hypercalciuria. Idiopathic hypercalciuria is frequently associated with nephrolithiasis; its relationship to nephrocalcinosis is less clear. In an interesting family reported by Buckalew et al. [20] individuals were identified with RTA, hypercalciuria and nephrocalcinosis, while other family members had hypercalciuria with nephrolithiasis (without nephrocalcinosis) and others had only hypercalciuria. These authors suggested that idiopathic hypercalciuria might lead to nephrocalcinosis and to RTA; however, this sequence of events appears to be rare, since most patients with hypercalciuria have neither nephrocalcinosis nor RTA.

f) Papillary Necrosis. This condition is particularly associated with analgesic abuse, and with pyelonephritis in diabetics. In abusers of analgesics, particularly those containing phenacetin, caliectasis may be seen as well as retained, sloughed, calcified papillae. The calcification is presumably mainly dystrophic, but may be exacerbated by the frequently associated RTA–1 and hypocitraturia. The papillae tend to be rather diffusely and uniformly calcified in this disorder.

g) Acetazolamide. This drug, now mainly used to treat glaucoma, causes a proximal type of RTA which is associated with hypocitraturia. Nephrocalcinosis occurs in experimental animals given the drug [21]; in patients receiving acetazolamide, nephrolithiasis is more frequent, and appears to be particularly associated with preexisting hypercalciuria [22].

h) Other Causes of Nephrocalcinosis. In vitamin D intoxication, sarcoidosis and idiopathic hypercalcemia of infancy, the nephrocalcinosis is presumably attributable to hypercalciuria and/or hypercalcemia. In hereditary hyperoxaluria, and in severe enteric hyperoxaluria, renal failure may be associated with calculi, and with medullary and cortical deposition of calcium oxalate. Other rare forms of hereditary nephrocalcinosis include that associated with renal magnesium wasting [23].

Radiologically demonstrable medullary nephrocalcinosis may show improvement when the underlying cause is treated, for example hyperparathyroidism, though rarely in less than two years [6]. The progression of nephrocalcinosis to end-stage renal failure is uncommon and is most often associated with hyperoxaluria and papillary necrosis. Although alkali treatment usually does not result in the disappearance of nephrocalcinosis in RTA–1, these patients rarely progress to end-stage renal disease. Bichler et al. [24] have shown that the calcium antagonist nifedipine may reduce the severity of nephrocalcinosis in an animal model; it is conceivable that similar treatment might affect nephrocalcinosis in man.

Comments

Nephrocalcinosis is a rather uncommon entity, which may be discovered accidentally, or because of symptoms usually due to associated nephrolithiasis. It may rarely provide a very useful diagnostic clue, for example in a patient presenting with hypokalemic paralysis due to renal potassium wasting secondary to RTA-1 in whom an abdominal X-ray may immediately confirm the diagnosis by showing nephrocalcinosis. Appropriate investigation permits a specific etiological diagnosis to be made in almost all cases. In some underlying disorders, treatment is available which should prevent progression of the nephrocalcinosis and end-stage renal disease; for example parathyroidectomy for underlying primary hyperparathyroidism, sodium or potassium bicarbonate or citrate for RTA-1, and withdrawal of the offending agents in cases of calcium or vitamin D excess, analgesic abuse or acetazolamide nephrocalcinosis.

References

1. Albright F, Baird PC, Cope O, Bloomberg E (1934) Studies on the physiology of the parathyroid glands. IV. Renal complications of hyperparathyroidism. Am J Med Sci 187:49–65
2. Alon U, Brewer WH, Chan ACM (1983) Nephrocalcinosis: detection by ultrasonography. Pediatrics 71:970–973
3. Anderson CK (1976) Pathology of nephrocalcinosis and stone formation. In: Williams DI, Chisholm GT (eds) Scientific foundations of urology, chap 4. Heinemann, London, pp 282–289
4. Randall A (1937) The origin and growth of renal calculi. Ann Surg 105:1009–1027
5. Wrong OM, Feest TG (1976) Nephrocalcinosis. In: Peters DK (ed) Advanced medicine, vol 12. Tumbridge Wells, Kent, Pitman Medical, pp 394–406
6. Wrong O (1985) The significance of radiological nephrocalcinosis. Hospital Update, pp 167–178
7. Lloyd-Thomas HG, Balme RH, Key JJ (1962) Tramline calcification in renal cortical necrosis. Br Med J 1:909–911
8. Broadus AE, Horst RL, Lang R, Littledike ET, Rasmussen H (1980) The importance of circulating 1,25-dihydroxyvitamin D in the pathogenesis of hypercalciuria and renal stone formation in primary hyperparathyroidism. N Engl J Med 302:421–426
9. Pak CYC, Nicar MJ, Peterson R, Zerwekh JE, Snider W (1981) A lack of unique pathophysiologic backgrounds for nephrolithiasis of primary hyperparathyroidism. J Clin Endocrinol Metab 53:536–542
10. Brenner RJ, Spring DB, Sebastian A, McSherry EM, Gennant HK, Palubinska J, Morris RC Jr (1982) Incidence of radiographically evident bone disease, nephrocalcinosis and nephrolithiasis in various types of renal tubular acidosis. N Engl J Med 307:217–221
11. Wrong O, Davies HEF (1959) The excretion of acid in renal disease. Quarterly J Med 28:259–313
12. Harrington TM, Bunch TW, van den Berg CJ (1983) Renal tubular acidosis, a new look at treatment of musculoskeletal and renal disease. Mayo Clinic Proc 58:354–360
13. Sutton RAL, Wong NLM, Dirks JH (1979) Effects of metabolic acidosis and alkalosis on sodium and calcium transport in the dog kidney. Kidney Int 15:520–533

14. Batlle D, Itsarayounyuen K, Hayes S, Arruda JAL, Kurtzman NA (1982) Parathyroid hormone is not anticalciuric during chronic metabolic acidosis. Kidney Int 22:264–271
15. Coe FL, Parks JH (1980) Stone disease in hereditary distal renal tubular acidosis. Ann Int Med 93:60–61
16. Konnak JW, Kogan BA, Lau K (1982) Renal calculi associated with incomplete distal renal tubular acidosis. J Urol 128:900–902
17. Yendt ER, Jarzylo S, Finnis WA, Cohanim M (1981) Medullary sponge kidney (tubular ectasia). In: Smith LH, Robertson WG, Finlayson B (eds) Calcium urolithiasis. Urolithiasis: clinical and basic research. Plenum Press, New York, pp 105–112
18. Backman U, Danielson BG, Fellstrom B, Johansson G, Ljunghall S, Wilstrom B (1981) Clinical and laboratory findings in patients with medullary sponge kidney. In: Smith LH, Robertson WG, Finlayson B (eds) Urolithiasis: clinical and basic research. Plenum Press, New York, pp 113–120
19. Parks JH, Coe FL, Strauss AL (1982) Calcium nephrolithiasis and medullary sponge kidney in women. N Engl J Med 306:1088–1091
20. Buckalew VM Jr, Purvis ML, Shulman MG, Herndon CN, Rudman D (1974) Hereditary renal tubular acidosis. Medicine, Baltimore 53:229–254
21. Harrison HE, Harrison HC (1955) Inhibition of urine citrate and the production of renal calcinosis in the rat by acetazolamide (diamox) administration. J Clin Invest 34:1662–1670
22. Sutton RAL, Dewar J, Walker VR, Drance S (1983) Renal calculi and acetazolamide therapy. Abstracts, 15th Ann Mtg Am Soc Nephrol 44A
23. Evans RA, Carter JN, George CRP, Wolse RS, Newland RC, McDonald GD, Lawrence JR (1981) The congenital "magnesium losing kidney" – report of two patients. Quarterly J Med (1981), 50:39–52
24. Bichler KH, Strohmaier WL, Schanz F, Nelde HJ, Gaiser I, Schulze E, Schreiber M (1985) Zur Wirkung von Kalziumantagonisten (Nifedipin) auf die Nephrokalzinose und Kalziumausscheidung der Ratte. Urol Int 40:13–21

Nephrocalcinosis in the Kidney of the Rat on Atherogenic Diet and the Effect of Calcium Antagonists (Nifedipine)

H.-J. NELDE[1], K.-H. BICHLER[1], W. L. STROHMAIER[1], and W. KRIZ[2]

Introduction

Nephrocalcinosis, the sedimentation of calciferous deposits in renal tissue, has been described by several authors. There is contradictory evidence on the localization of concrements found, indicating that the pathogenesis is to a large extent unknown and that renal lesions present themselves in several ways. Probably these lesions are partly dystrophic calcifications caused by cell damage, as they are known to occur in other organs [6]; for example myocardnecrosis leads to calcification of heart-muscle cells. Experimental investigations demonstrated that these calcifications can be reduced by calcium antagonists [7, 10, 17]. Similar starting points result from experimental feeding of animals in which a nephrocalcinosis was induced. The aim of these investigations was to establish animal models to examine and explain alterations in the parameters which cause nephrocalcinosis.

In investigations of these questions the rat has proved to be of value. A nephrocalcinosis in the rat can be induced within a few weeks by feeding an atherogenic diet rich in cholesterol [13]. We wished to test whether calcium antagonists could have an effect on experimentally induced nephrocalcinosis, the renal processing of calcium and other parameters relevant in kidney stone formation [4, 11, 12, 14, 15].

In the following we try to visualize the extent of experimentally induced nephrocalcinosis both by light and electron microscopy.

Materials and Methods

Male Wistar rats (Chbb: THOM) were fed an atherogenous diet (C 1014; Altromin, Lage, FRG); as controls others received normocaloric food (C 1000; Altromin, Lage, FRG). As a calcium antagonist, nifedipine was administered in a daily dosage of 50 mg/kg body weight in two half doses through a stomach tube.

Apart from pathohistological alterations in rat kidney, a number of urine parameters which are said to be important for lithogenesis were also examined [4,

---

[1] Department of Urology, University of Tübingen, Calwer Str. 7, D-7400 Tübingen, FRG.
[2] Institute of Anatomy, University of Heidelberg, D-6900 Heidelberg, FRG.

Nephrocalcinosis, Calcium Antagonists,
and Kidney
Ed. by K.-H. Bichler and W.L. Strohmaier
© Springer-Verlag Berlin Heidelberg 1988

collectives:

group 1	atherogenous diet	n = 20
group 2	atherogenous diet + Nifedipine from the start of the experiment	n = 20
group 3	atherogenous diet + Nifedipine 4 weeks after the start of the experiment	n = 20
group 4	control diet	n = 10
group 5	control diet + Nifedipine	n = 10

Fig. 1. Examination material

15]. The animals were kept in metabolism cages and urine was pooled in weekly portions and deep-frozen. The experiment lasted 10 weeks.

A total of 80 male Wistar rats were arbitrarily divided into five groups: groups 1 to 3 were given an atherogenous diet, group 2 an atherogenous diet plus nifedipine, group 3 nifedipine 4 weeks after the start of the experiment, group 4 and 5 were fed on a control diet, group 5 receiving nifedipine in addition (Fig. 1).

At the end the animals were prepared for tissue collection. Paraffin sections made for light microscopic examinations did not always allow for an exact localization of the concrements. As the topography seemed to be of great importance in explaining the origin of nephrocalcinosis, electron microscopic examinations were carried out as well. For this purpose a perfusion fixation was carried out via the renal artery in anaesthetized rats ($n = 10$), (0.2 ml Inactin/100 g body weight; Byk-Gulden, Konstanz, FRG).

The fixation of kidney tissue was done with a perfusion machine using a constant perfusion pressure of 200 mmHg via the aorta abdominalis behind the exit of the arteria renalis. First the kidneys were washed with about 200 ml of an irrigation fluid, followed by a fixation solution [8]. The solution was 3% glutaraldehyde and saccharine was used to give an osmolarity of 600 mosmol. Perfusion was correct when an increase in tissue consistency and a change in tissue color was seen (picrin acid!). The kidneys were then removed, embedded in synthetic material (plastics) and prepared for the electron microscopic examination [8].

We performed an autopsy on other rats and examined the kidneys or efferent urinary tract for macroscopically visible concrements then removed them for further histological examinations. Tissue fixation was made in buffered formaline (4%, pH 7) and histological section material was embedded in paraffin wax for examination. The localization and quantification of the concrements was done by a coloring with naphthalhydroxamic acid using Voight's method [16]. The concrement calcification became clearly visible in polarized light as a double refraction.

The extent of calcification was graded semiquantitatively into groups ranging from "no calcification" to "massive calcification" ($-$, $+$, $++$, $+++$) (Fig. 2). The analysis was carried out by four independent coworkers who did not know to which group the rats belonged or what the analysis of their coworkers was. The average calcification index was calculated for each group using the predetermined calcification grades. Furthermore a concrement analysis was done using the polarizing microscope and X-ray diffraction to determine the qualitative and quantitative composition.

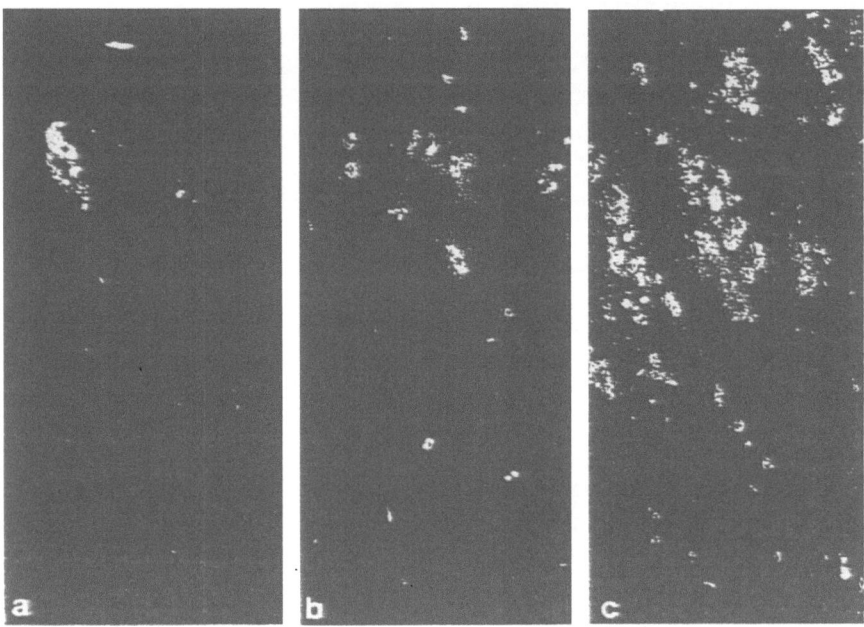

Fig. 2 a–c. Quantification of calcification by polarizing microscopy (grading of calcification: **a** Low, **b** moderate, **c** strong)

In parallel sections an immunohistological examination using Tamm-Horsfall protein (TH protein, uromucoid) was done. An immunoperoxidase method with the avidin-biotin system (ABC-kit, Vectastain Labs, Inc.: Burlingame, USA) and a rabbit antirat serum as primary antibody was used [2, 3, 9].

It was possible to correlate the quantity of calcification of $n = 17$ animals of groups 1–3 and $n = 15$ animals from groups 4 and 5 with the excretion of TH protein in urine. The determination of TH protein was done by rocket immunodiffusion according to Laurell [described in 2,3].

Results

Our examination demonstrated:

1. Except for isolated concrements there was no calcification in the cortex and medullary zone of the kidney control rats (Fig. 3).
2. However, there was a more or less clear nephrocalcinosis to be seen in rats on the atherogenic diet (Fig. 4 a–c). Seen under the light microscope the deposits were located in the corticomedullary transitional zones. The arrangement of concrements partially in radial direction might indicate a relation with both the course of the nephron and the vessel. In the cortex zone there was no recognizable calcification. Numerous concrements, however, were found in the inner and outer stripe of the exterior zone of the medulla and in a few isolated cases also in the inner medullary zone and the incised renal pelvis.

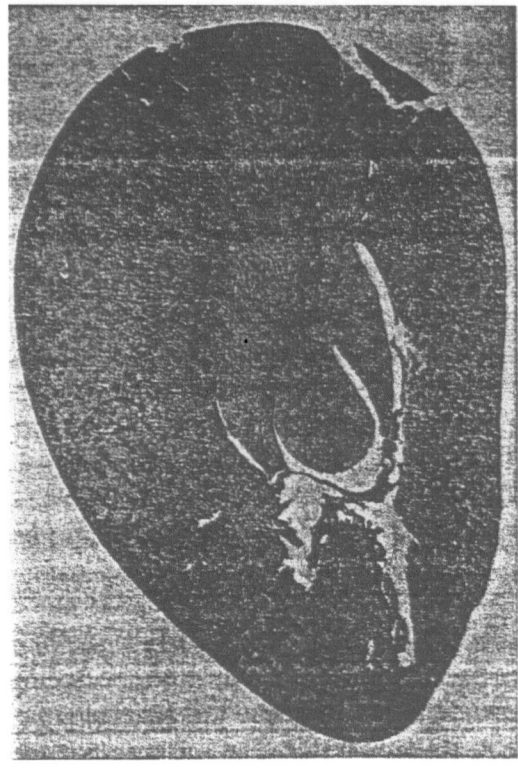

Fig. 3. Control rat, histological cross-section

Figure 5 shows the differences in the calcification grade for each group. On the abscissa are the individual results, on the ordinate the number of animals under examination. It can be seen that animals on a control diet or a control diet plus nifedipine had only slight calcification. The calcification index was 0.2 or 0.6. In the groups on an atherogenous diet, however, the increased calcification due to a shift towards a higher calcification grade is noticeable. The index here was 2.35 in group 1 and 2.45 in group 3. It is striking that the animals in group 2 – that is on an atherogenous diet plus nifedipine from the beginning of the experiment – showed markedly fewer concrement formations, with a calcification index of 1.67.

Under the optical microscope the concrements displayed a concentric formation of layers which were partially granularly disintegrated. The analysis of the concrements for possible stone forming substances showed calcium phosphate and struvite crystals. With the electron microscopic examination two different types of deposit were found: intraluminal and interstitial calcification.

Intraluminally only a few calcifications were found. Morphologically they showed a structure of crystal needles, concentrated as a conglomerate (Fig. 6 a, b). The figure shows such a concrement in the lumen of the distal tubule. Most deposits were found in the interstitium (Fig. 6 a). These are markedly different morphologically from those described before. A concentric layer indicates growth by apposition in these concrements.

Immunological examination of TH protein demonstrated a marked reaction in the distal tubules of the control group (Fig. 7 a). The identification product was

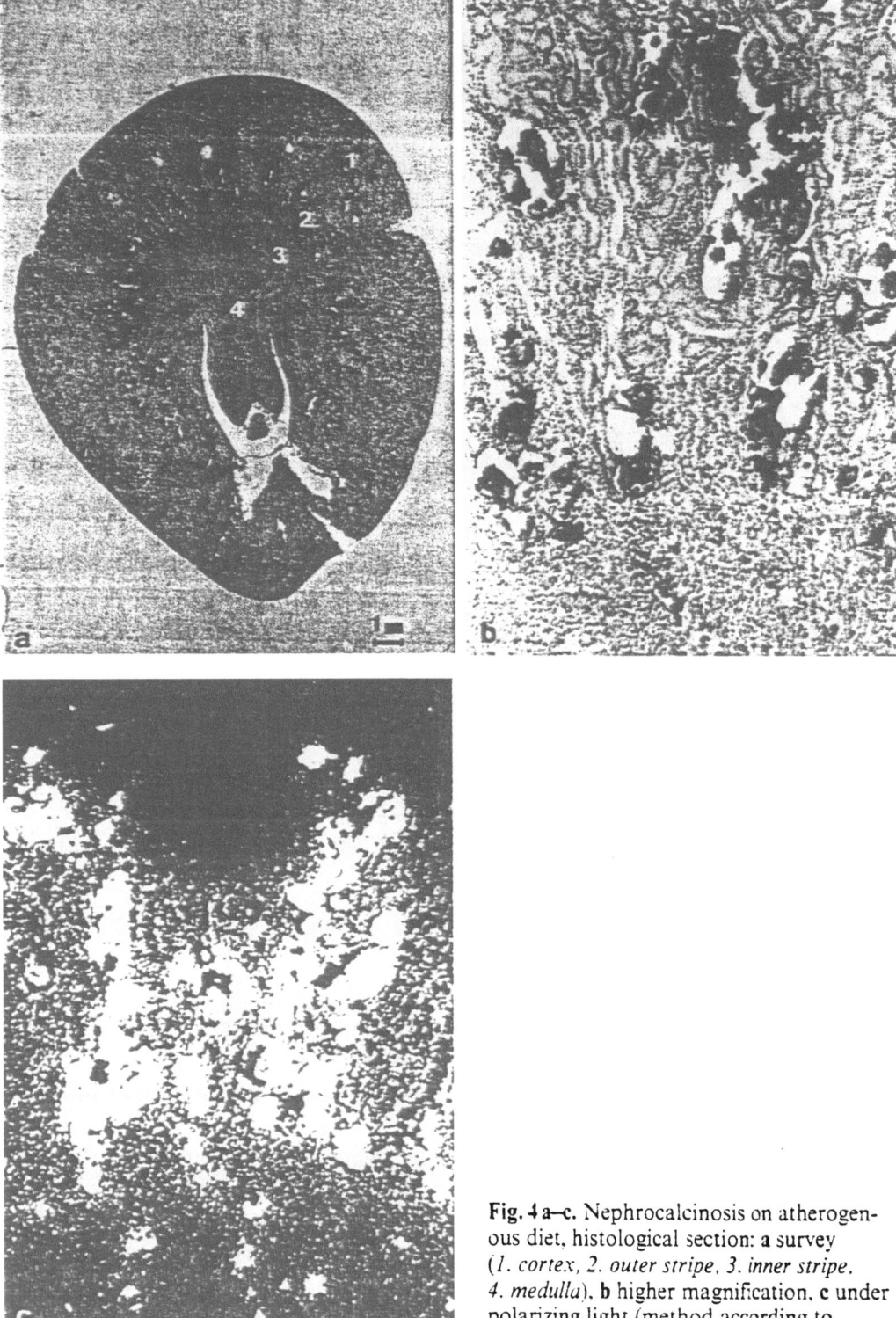

Fig. 4 a–c. Nephrocalcinosis on atherogenous diet, histological section: **a** survey
(*1. cortex, 2. outer stripe, 3. inner stripe, 4. medulla*), **b** higher magnification, **c** under polarizing light (method according to Voight)

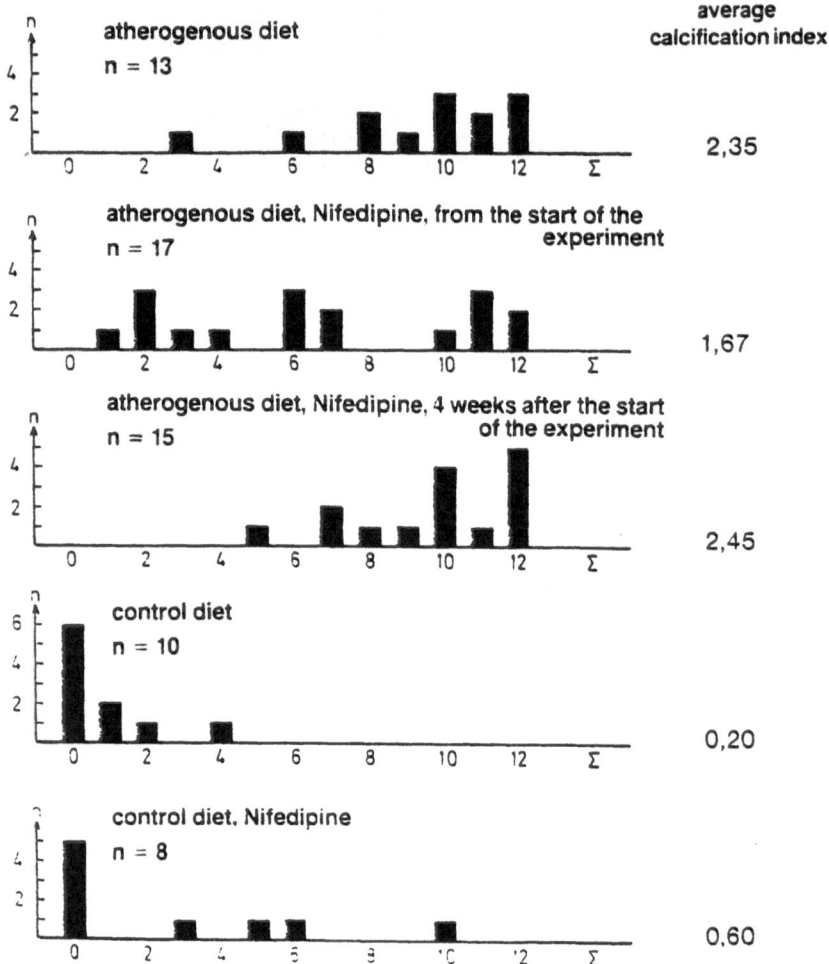

Fig. 5. Quantification of the extent of calcification

located intracellularly in the cytoplasm. An intraluminal reaction with a marked marginal intensification in the area of the luminal cell membrane was seen. The kidneys of animals on an atherogenous diet showed only a slight immune reaction in the same areas, indicative of reduced production and secretion of uromucoid (Fig. 7b).

From the second week onwards the atherogenous diet led to an extreme decrease of uromucoid excretion in the urine (Fig. 8) which lasted till the end of the experiment. This reduction occurs with an increase of deposits in the tissue. The correlation between TH protein excretion in the urine during the last collecting period and the corresponding renal calcification is shown in Fig. 9. In control rats with normal excretion of uromucoid there were only smaller deposits. On the contrary the significantly lowered excretion of uromucoid in the urine agrees with an increase in calcification. Using the immunohistological identification reaction of TH protein, this urinary protein could be partly found in the deposits.

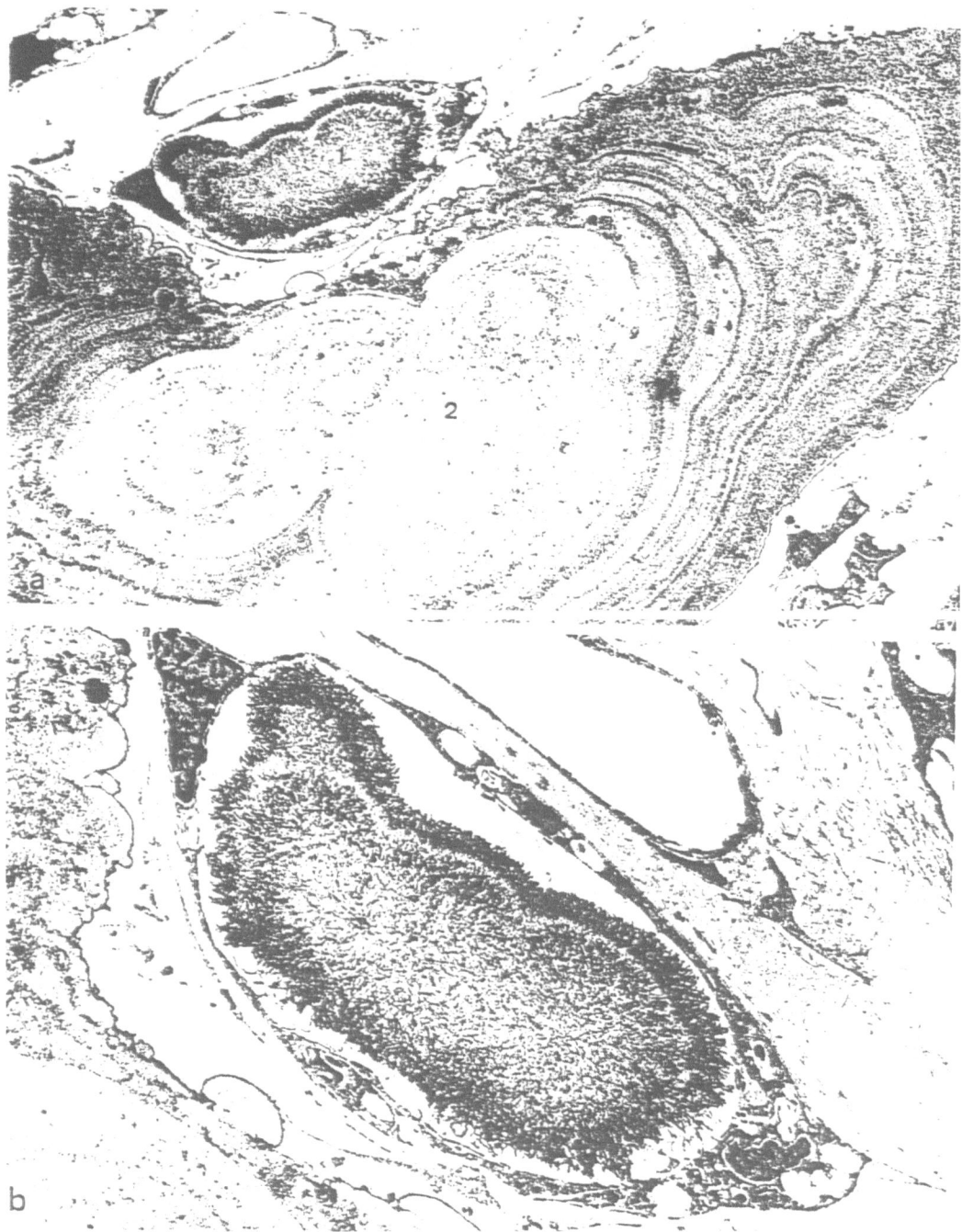

Fig. 6 a, b. Transmission electron microscopic examination of calcification. a Survey (*1. intraluminal*, *2. interstitial*), b intraluminal deposit, higher magnification

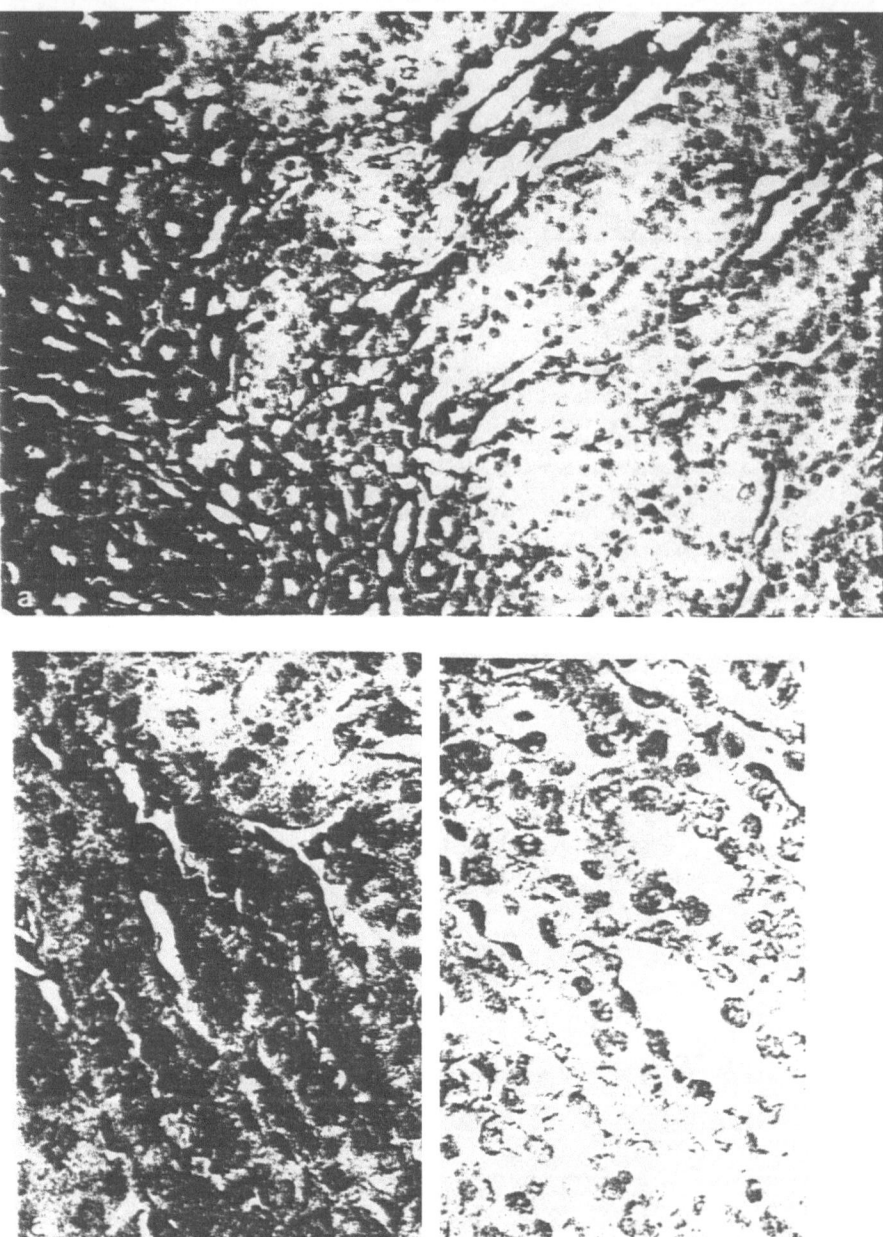

Fig. 7 a, b. Immunohistological localization of *TH protein* (uromucoid) in the rat kidney: **a** control rat, **b** under atherogenous diet

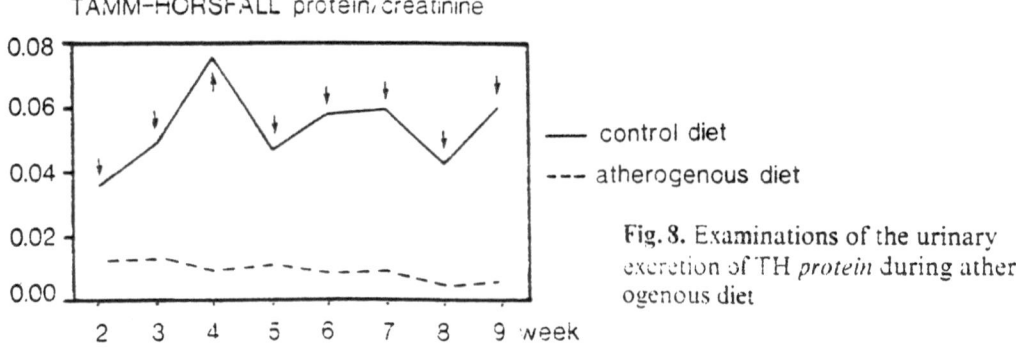

Fig. 8. Examinations of the urinary excretion of TH *protein* during atherogenous diet

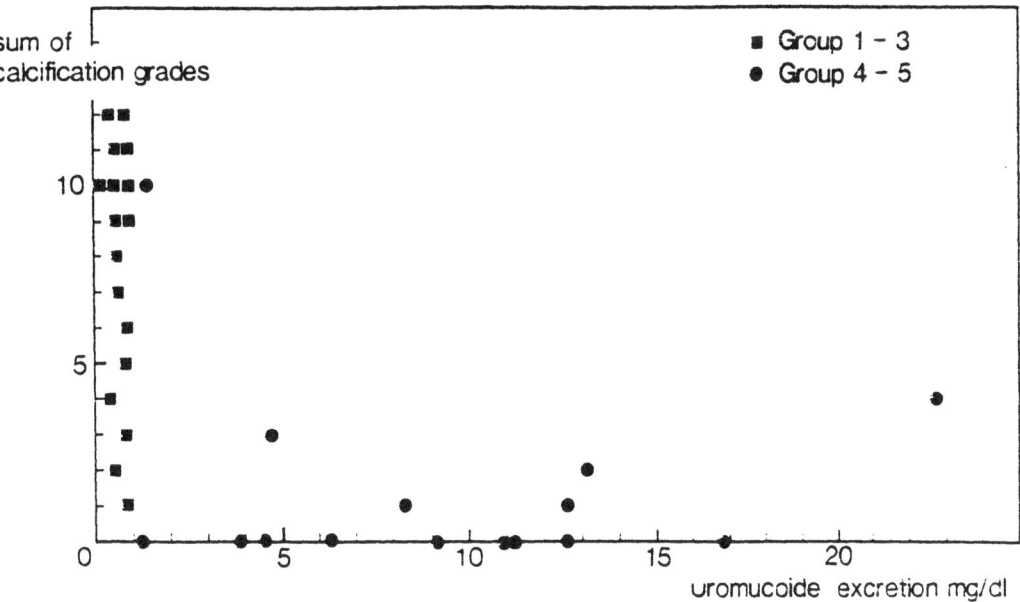

Fig. 9. Correlation of *TH protein* excretion in urine versus grading of calcification

Discussion

What are the conclusions to be drawn from these examinations of the pathogenesis of nephrocalcinosis? On one hand we have to consider whether there are two totally different developmental mechanisms, an indication for this being the different morphologies.

In principal there are three places for localization of nephrocalcinotic deposits: intraluminally, interstitially, and intracellularly. For intraluminal development of concrements as precursors of stone formation we must consider alterations in urine composition, e.g., pH alterations and altered secretion of so-called inhibitory substances, that are a result of physicochemical processes and which would mean that the solubility product of certain ions is exceeded. Besides this, intraluminal calcification may primarily develop intracellularly. Also for interstitially positioned deposits, the question is whether they develop primarily, in the interstitium, or secondarily, as a result of defects in tubule cells causing leakage of different substances into the interstitium. For example, Caulfield describes concentrically stratified deposits after application of calcium gluconate (Fig. 10) similar to those we saw. He found some indications that the tubule cells are lifted off the basal membrane by these deposits and that gradual decrease in function finally leads to cell death. Furthermore calcification which started primarily in the mitochondria of the tubular cells developed after the administration of parathyroid hormone (PTH) (Fig. 11). These mitochondria seem to stop working, which results in further crystal accumulation in the cytoplasm. In a terminal state this leads to cell wall rupture and a prolapse of calcification into the urinary canal lumen.

It is possible that the calcification we have described developed primarily intracellularly, were set free by tubule cell death, and thus only seem to be located interstitially.

3 days

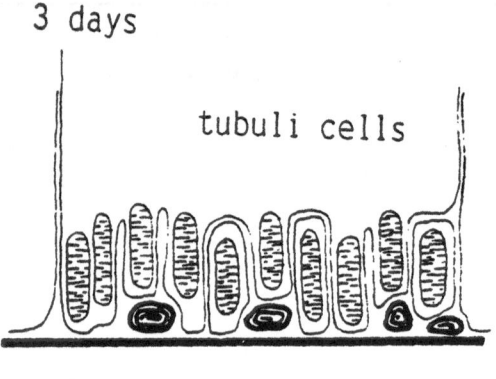

tubuli cells

18 days

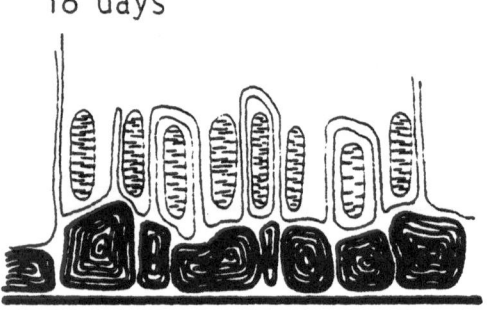

interstitial deposits

Fig. 10. Induction of nephrocalcinosis by giving calcium gluconate (according to Caulfield)

50^h after PTH

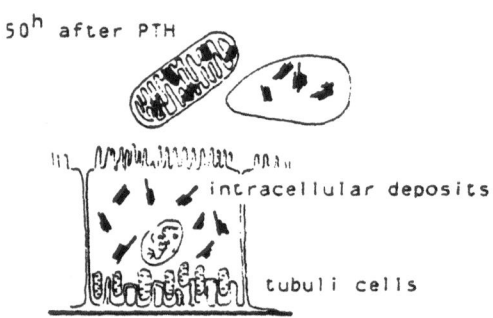

intracellular deposits

tubuli cells

60^h after PTH

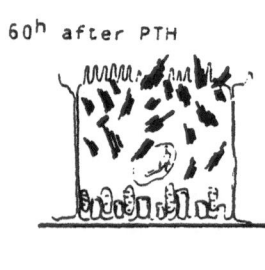

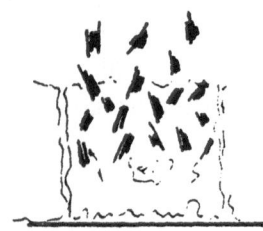

Fig. 11. Induction of nephrocalcinosis by giving *PTH* (according to Caulfield)

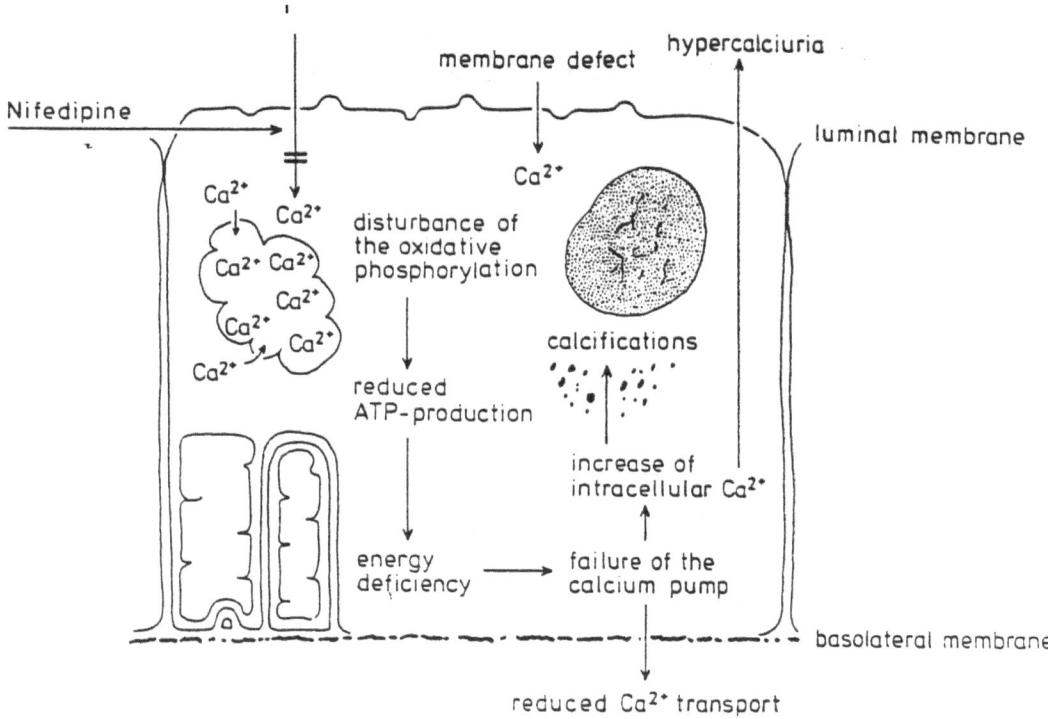

Fig. 12. Hypothesis of calcification reduction with calcium antagonists

In this context we must consider the cellular alteration of efferent urinary canals and look at the first pathophysiologic alterations in the substructure of tubule cells. Primary defects are obviously intracellular; various authors discuss defects in energy methabolism. In this context examinations of TH protein excretion would make sense as well. As TH protein is a product of distal tubule cells, an alteration of TH protein excretion indicates a cellular alteration in this area. As TH protein is produced in the tubule cells of the distal tubules [9] and is then excreted into the lumen of the urinary canals, in general it can only be identified in concrements which are located in the lumena. If uromucoid is deposited in concrements which from their location originated in the interstitium, the damage of the interstitium has to be regarded as the cause. This can be interpreted as an indication of cell defect. Summarizing these results we can offer the following hypothesis (Fig. 12). Interstitial and intraluminal calcifications most probably originate primarily in intracellular processes (as has been described by Caulfield and as far as the results we obtained indicate). It might be assumed that an atherogenous diet leads to membrane defects via metabolic disturbances particularly in the distal areas of the nephron which are ontogenetically highly differentiated. The inflow of calcium into the cell resulting from these defects with calcium overloading and damage of the mitochondria causes disturbances in the energy balance and subsequent failure of the calcium pump mechanisms in the basolateral membrane. An increase in intracellular calcium results in overloading of the cell and calcification. We think support for this hypothesis is the mitochondrial calcification resulting from cell overloading with calcium as Caulfield found. This opinion is supported by the fact that calcium antagonists have an inhibitory effect on nephrocalcinosis. The effect of calcium antagonists in combination with an atherogenous diet might be reduction or prevention of the influx of calcium into the altered cell. This might be indicated by the fact that calcium antagonists which primarily act intracellularly – as can be demonstrated for the heart-muscle cell – reduced the calcification.

We can only speculate on the genesis of the deposits and particularly on the formations in this area. On the other hand the morphology of the kidney on the borderline of macula and cortex is the critical point. The venous blood of the medulla flows via the vasa recta and in the outer strip the whole venous blood flows in wide luminal vessels between the tubules to end in the vena arcuata localized in the medulla-cortex borderline. It is possible that there are disturbances of the outflow in this area, a slowing down of the blood flow and then deposit formation. With growth in deposit size the resistance to venous blood flow increases and a continuous growth of the deposits causes an inhibition of the urine outflow and crystallization.

Could there be an improved regulation of the blood stream in the above-mentioned area of the kidney by calcium antagonists? To answer this we must consider whether the mechanisms of the effect of calcium antagonists start in the blood supply. Abe et al. [1] described a switch-over in the corticomedullary zone of the kidney induced by calcium antagonists. By that means different areas can be supplied with blood much better, leading to a reduction of deficient metabolic processes and hence to a reduction of concrements.

Thus the aim of further investigations should be the study both morphologically and physiologically of cellular alterations.

References

1. Abe Y, Komori T, Miura K, Takada T, Imaniski M, Okahara T, Yamamoto K (1983) Effects of the calcium antagonist nicardipine on renal function and renin release in dogs. J Cardiovascular Pharmacol 5:254–259
2. Bichler KH, Haupt H, Uhlemann G, Schwick HG (1973) Human uromucoid. I. Quantitative immunoassay. Urol Res 1:50–59
3. Bichler KH, Kirchner C, Ideler V (1976) Uromucoid excretion of normal individuals and stone formers. Br J Urol 47:733–738
4. Bichler KH, Strohmaier WI, Schanz F, Nelde HJ, Gaiser I, Schulze E, Schreiber M (1985) Zur Wirkung von Kalziumantagonisten (Nifedipin) auf die Nephrokalzinose und Kalziumausscheidung der Ratte. Urol Int 40:13–21
5. Caulfield JB, Schrag PE (1964) Electron microscopic study of renal calcification. Am J Pathol 44:365–381
6. Gedigk P, Totovic V (1977) Zell- und Gewebsschäden. In: Eder M, Gedigk P (Hrsg) Lehrbuch der Allgemeinen Pathologie und der Pathologischen Anatomie. Springer, Berlin Heidelberg New York, S 1–68
7. Jamison RL, Kriz W (1982) Urinary concentrating mechanism: structure and function. Oxford University Press, New York
8. Kaissling B, Kriz W (1982) Variability of intercellular spaces between macula densa cells: a TEM study rabbits and rats. Kidney Int 22:9–17
9. Kirchner C, Bichler KH (1976) Uromucoid in the rat: its isolation, localization in the kidney and concentration in the urine. Urol Res 4:119–123

10. Lossnitzer K, Janke J, Hein B, Stauch M, Fleckenstein A (1975) Disturbed myocardial calcium metabolism: a possible pathogenetic factor in the hereditary cardiomyopathy of the Syrian hamster. In: Fleckenstein A, Rona (eds) Recent advances in studies on cardiac structure and metabolism, vol 6. University Park Press, Baltimore, pp 207–216
11. Nelde HJ, Bichler KH, Strohmaier WL (1984) Histopathological changes of the rat kidney under atherogenous diet (abstract). 5th Int Symp on Urolithiasis and Related Research, Garmisch-Partenkirchen 1983. Urol Res 12:57
12. Nelde HJ, Bichler KH, Strohmaier WL, Schanz F (1985) Histopathologische Veränderungen in der Niere der Ratte unter atherogener Diät bzw. Nifedipin-Verabreichung. In: Gasser G, Vahlensieck W (Hrsg) Pathogenese und Klinik der Harnsteine. XI. Fortschritte der Urologie und Nephrologie 23. Steinkopff, Darmstadt, S 319–325
13. Schwille PO, Brandt P, Ulbrich D, Kömpf W (1975) Pankreasinseln, Plasmaglucagon und renale Verkalkungen unter verschiedener Grunddiät bei der Ratte. Urologe A 14:306–314
14. Strohmaier WL, Bichler KH, Gaiser I, Schulze E, Nelde HJ, Schreiber M (1984) Urine and serum levels related to lithogenesis in rats on an atherogenous diet (abstract). 5th Int Symp on Urolithiasis and Related Clinical Research, Garmisch-Partenkirchen 1983. Urol Res 12:57
15. Strohmaier WL, Bichler KH, Schanz F, Nelde HJ, Schreiber M (1985) Einfluß von Kalziumantagonisten auf die Ausscheidung von Kalzium und anderer für die Steingenese wichtiger Substanzen bei der Ratte. In: Gasser G, Vahlensieck W (Hrsg) Pathogenese und Klinik der Harnsteine. XI. Fortschritte der Urologie und Nephrologie 23. Steinkopff, Darmstadt, S 365–372
16. Voigt GE (1957) Ein neuer histochemischer Nachweis des Calciums (mit Naphtalhydroxamsäure). Acta histochem 4:122–131
17. Wrogemann K, Pena SD (1976) Mitochondrial calcium overload; a general mechanism for cell necroses in muscle diseases. Lancet I:672–673

Influence of Calcium Antagonists (Nifedipine) on the Excretion of Calcium and Other Substances Relevant for Stone Formation in Rats with Nephrocalcinosis (Induced by Atherogenic Diet)

W. L. Strohmaier[1] and K.-H. Bichler[1]

Introduction

Until recently nephrocalcinosis could not be treated very well by therapy and eventually resulted in renal insufficiency. The pathogenesis of this disease is virtually unknown but pathologically we can distinguish between a metastatic and dystrophic type of calcification. The latter is caused by cell damage as is known to happen in other organs. For example, necrosis of the myocardium leads to calcification of heart muscle cells. In experimental investigations it was demonstrated that these calcifications could be reduced by calcium antagonists [8, 13, 24].

The following investigation was designed to test whether the calcium antagonist nifedipine could have an effect on experimentally induced nephrocalcinosis or the renal handling of calcium and other parameters related to the pathogenesis of nephrocalcinosis and calcium urolithiasis.

Methods

For this purpose we used an animal model as described before [4, 20]. By feeding an atherogenic diet (rich in cholesterol, 10 g/kg, Altromin Lage/Lippe, FRG) a distinct nephrocalcinosis occurred readily after 4 weeks [17].

A total of 80 male rats (CHBB: Thom) were arbitrarily divided into 5 groups:
group 1: atherogenic diet
group 2: atherogenic diet plus calcium antagonist
group 3: atherogenic diet plus calcium antagonist 4 weeks after the beginning of the experiment
group 4: control diet
group 5: control diet plus calcium antagonist

As calcium antagonist we gave nifedipine in a daily dosage of 50 mg/kg body weight through an esophagal tube in two individual doses. The animals were kept in metabolism cages, the urine was collected for 8 h a day and pooled into weekly collections. After 10 weeks the experiment was ended and the animals were killed and dissected [4].

---

[1] Department of Urology, University of Tübingen, Calwer Str. 7, D-7400 Tübingen, FRG.

Nephrocalcinosis, Calcium Antagonists, and Kidney
Ed. by K.-H. Bichler and W.L. Strohmaier
© Springer-Verlag Berlin Heidelberg 1988

Apart from the histological and electron microscopic examinations, also described in this book (Nelde et al.), we examined a number of urine parameters related to urolithiasis: calcium, magnesium, and sodium were measured by flame photometry or atomic absorption respectively (FL 6, Zeiss/Oberkochen). Creatinine was measured by a Beckman analyzer, the method is based on the Jaffé reaction. Citrate was measured by a commercial test kit (Boehringer Mannheim GmbH, Mannheim, FRG), based on the citrate lyase reaction. The Tamm-Horsfall protein was determined by Laurell's electroimmuno diffusion as described in previous studies [3,-19]. For each parameter the creatinine ratio (x/creatinine) was calculated. For statistical analysis the Mann-Whitney U test was used [18].

Results

Analysis of the above parameters revealed the following: Atherogenic diet led to an increased calcium excretion in the urine. Compared with the results of an ear-

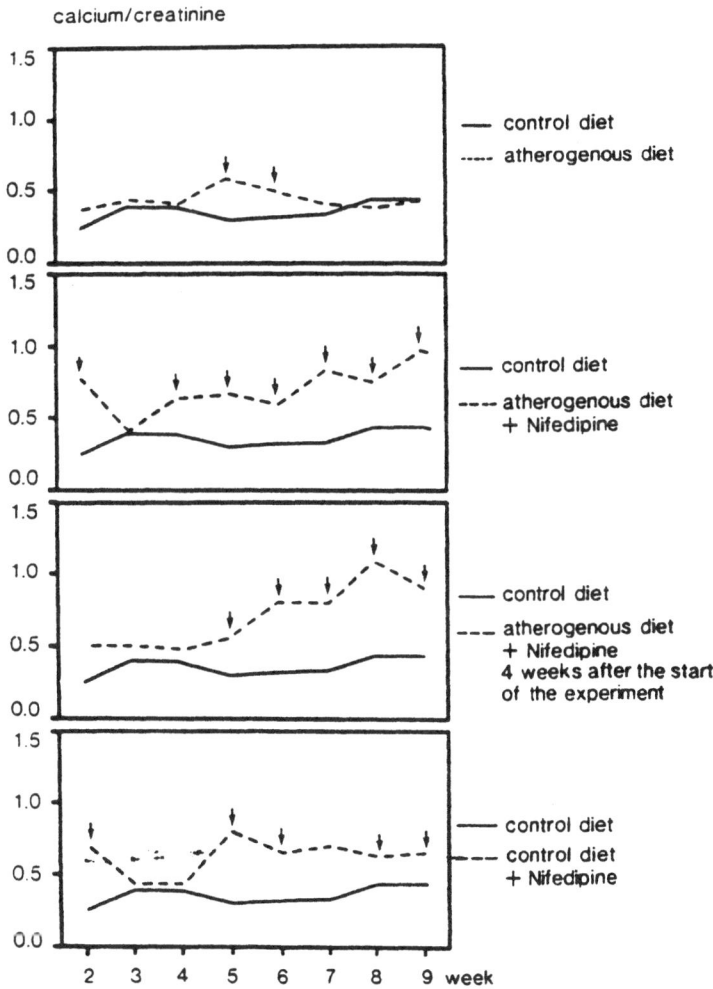

Fig. 1. Mean values of calcium excretion in the urine (calcium/creatinine, mmol/mmol). *Arrows* show significant differences ($p \leq 0.05$)

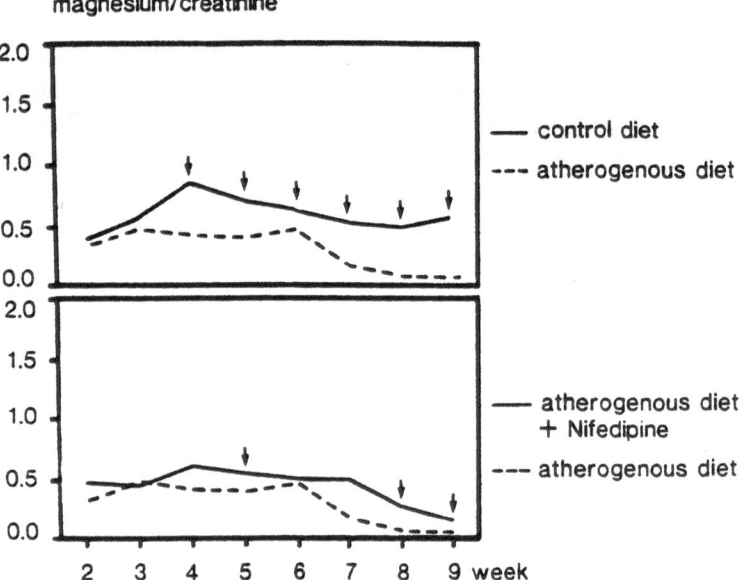

Fig. 2. Mean values of magnesium excretion in the urine (magnesium/creatinine, mmol/mmol). *Arrows* show significant differences ($p \leq 0.05$)

lier investigation using this animal model [20] the result was not very striking. Nifedipine clearly increased calcium excretion in combination with both atherogenic diet and the control diet. This was particularly noticeable in group 3 where the calcium excretion was increased immediately after nifedipine was given 4 weeks after the start of the experiment (Fig. 1).

With the atherogenic diet we found a lowered excretion of magnesium. With the combination of this diet plus nifedipine we found, however, a higher quantity of magnesium than with the atherogenic diet alone; that is nifedipine weakened the hypomagnesuria effect of the atherogenic diet (Fig. 2). The atherogenic diet did not change the sodium excretion in the urine; nifedipine, however, in combination with the control diet led to a significant sodium diuresis from the fifth week onwards (Fig. 3).

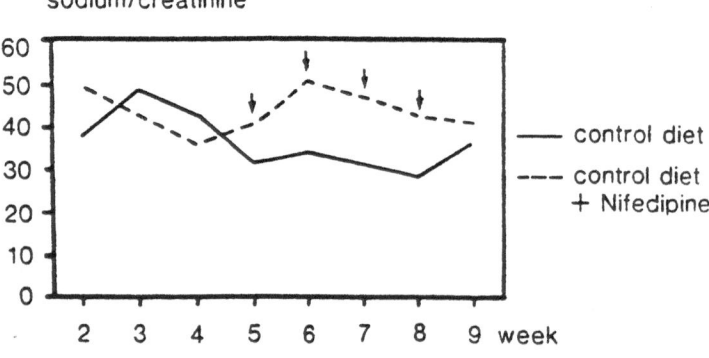

Fig. 3. Mean values of sodium excretion in the urine (sodium/creatinine, mmol/mmol). *Arrows* show significant differences ($p \leq 0.05$)

citrate/creatinine

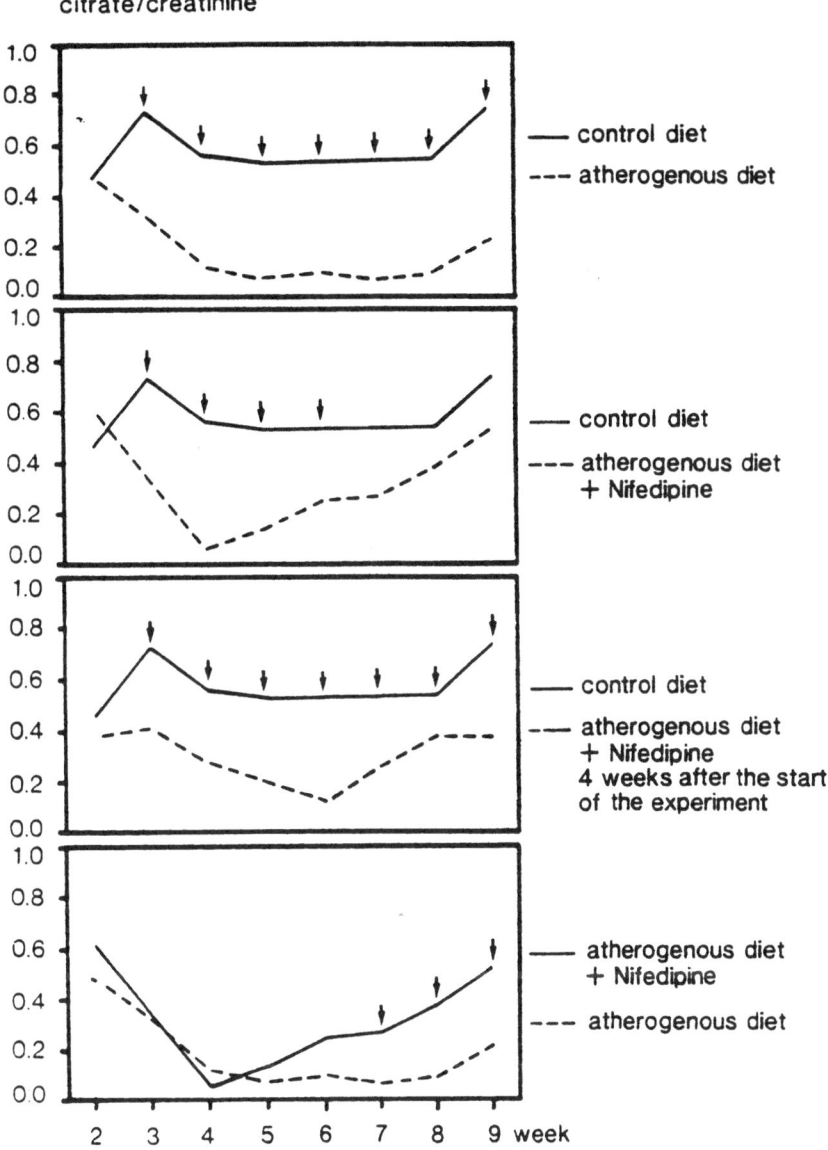

Fig. 4. Mean values of citrate excretion in the urine (citrate/creatinine, mmol/mmol). *Arrows* show significant differences ($p \leqq 0.05$)

The excretion of citrate was significantly lowered due to the atherogenic diet. Nifedipine brought about a noticeable increase in the quantity of citrate particularly in the last weeks of the experiment (Fig. 4).

Notably the atherogenic diet led to an extreme decrease in the excretion of Tamm-Horsfall protein in the second week of the experiment. The additional administration of nifedipine did not affect this development (Fig. 5).

Summarizing we found the following effects of nifedipine: it increased the excretion of calcium and sodium significantly and the decreasing effect of the atherogenic diet on the excretion of magnesium and citrate could be weakened.

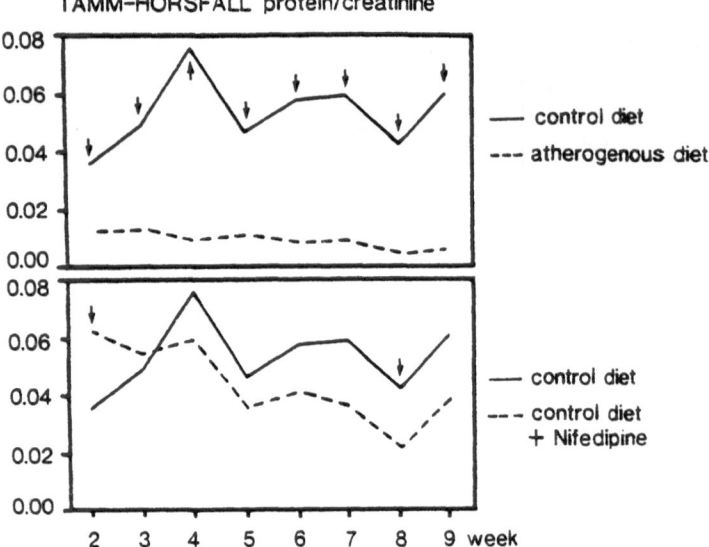

Fig. 5. Mean values of Tamm-Horsfall protein excretion in the urine (Tamm-Horsfall protein/creatinine, NE (arbitrary unit)/mmol). The arbitrary unit was defined as the amount of Tamm-Horsfall protein in 100 ml pooled urine of normal rats [18]. *Arrows* show significant differences ($p \leqq 0.05$)

Discussion

In order to analyze these effects of nifedipine it is necessary to consider several mechanisms. First the origin of calcium and sodium diuresis produced by calcium antagonists must be in the kidney itself, because the same effects can be seen in isolated perfused kidneys or following a direct injection of calcium antagonist into the renal artery [2, 15]. A number of renal mechanisms might be responsible for the increased diuresis, natriuresis, and calciuresis. They may be caused by an increased renal blood flow and glomerular filtration rate as was demonstrated by several authors after parenteral application of calcium antagonists [2, 14, 25]. The simultaneous increase of both renal blood flow and glomerular filtration rate seems to be a unique effect of calcium antagonists. In contrast most vasodilators increase the renal blood flow without raising the glomerular filtration rate. This unique effect of calcium antagonists may be due to preferential antagonism of the preglomerular vasoconstriction without decreasing the vascular tone in the efferent vessel [14].

After oral administration of calcium antagonists particularly, a significant increase in the filtration rate is not common [5, 12, 23]. Furthermore Yamaguchi [25] could demonstrate a natriuretic response even when the renal blood flow and the glomerular filtration rate were kept constant by constriction with an aortic clamp. Massry and Kleemann [16] also demonstrated that an increased glomerular filtration rate up to 85% caused only slight changes in the sodium and calcium excretion. Thus natriuresis and calciuresis with calcium antagonists cannot be explained by an increased glomerular filtration rate alone.

A decrease of the release of renin might also cause natriuresis. Many studies, however, showed a stimulatory effect of calcium antagonists on renin metabolism [11, 12, 15, 26]. Furthermore a suppression of aldosterone secretion has to be considered. However, several authors, e.g., Leonetti [12] and Lederballe [11], did not find a decrease in aldosterone following the administration of calcium antagonists.

Finally a direct action on the tubule cell handling of sodium and calcium should be discussed. Some arguments support this hypothesis. Sodium and calcium excretion with calcium antagonists occur in parallel as do the transport mechanisms of calcium and sodium at least in the proximal tubule [6, 21]. Wallia et al. [23] showed increased diuresis, and excretion of sodium, calcium, and phosphate on oral administration of nitrendipine to humans without any change of renal blood flow or glomerular filtration rate, suggesting a direct tubule cell action of calcium antagonists.

For this reason we now want to discuss a hypothesis for direct interaction between calcium antagonists and the tubule cell reabsorption mechanisms (Fig. 6). The calcium resorption mechanism of the proximal tubule is probably a sodium-calcium countertransport [22]. A sodium-potassium ATPase in the basolateral membrane pumps sodium from the cell into the interstitium; a resulting electrochemical gradient of sodium ions allows the influx of sodium from the tubule into the cell. At the same time calcium ions are forced out of the cell into the interstitium. This gradient brings about an influx of calcium from the tubule into the cell. In addition to the sodium-potassium ATPase a calcium sensitive ATPase of the basolateral membrane has been described [9]. However, it is still not possible to determine which of the two resorption mechanisms is of greater importance for calcium processing.

The following effect in the tubule cells might be attributed to nifedipine: calcium antagonists concentrate in the luminal membrane and thus hinder the influx

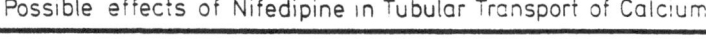

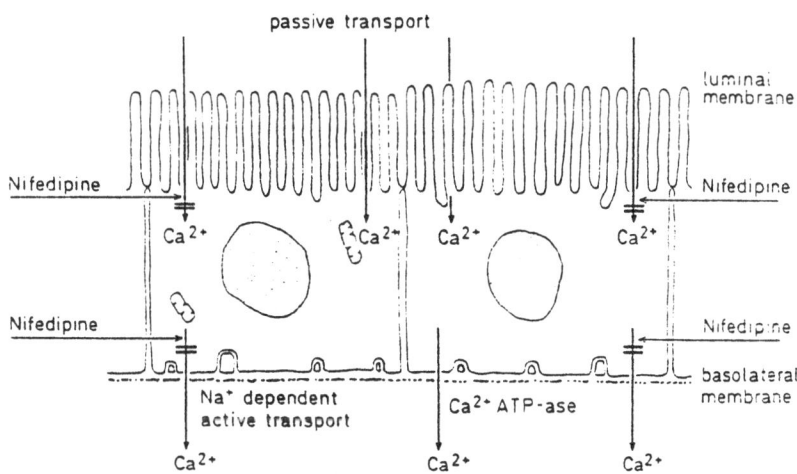

Fig. 6. Hypothetic action of calcium antagonist on calcium and sodium transport mechanisms in the proximal tubule. Explanations are given in the text

of calcium and sodium into the cell. Previous investigations [7] showed, however, that verapamil had no direct effect on calcium resorption if applied directly to the lumen of the proximal tubule. It is possible, however, that calcium antagonists cause a blockage of the calcium channel in the basolateral membrane, which impairs the outflow of calcium into the interstitium. As a result the intracellular calcium concentration is increased as is the number of positive ions. In this way a further influx of calcium and sodium into the tubule cell is inhibited, resulting in an increased excretion of sodium and calcium.

A recent investigation by Lang and Deetjen [10] demonstrated that calcium antagonists can block calcium channels in cultured MDCK cells which are similar to distal tubule cells. Thus this investigation might indicate a direct effect of calcium antagonists on tubule cell electrolyte transport mechanisms.

Finally it is noteworthy that nifedipine increased the excretion of so-called inhibitory substances significantly (e.g., citrate which was clearly decreased by the atherogenic diet). The reason for this remains unclear; it is possible that the reduced citrate excretion on an atherogenic diet is caused by acidosis of damaged tubule cells leading to an increased cellular citrate intake [1]. The intracellular acidosis could be reduced by a protective action of the calcium antagonist thus increasing the citrate excretion again. This might also be a contributing factor to the lower grade of renal calcification with nifedipine. It was particularly surprising to us that these calcifications were markedly reduced by nifedipine although this drug caused a significant hypercalciuria at the same time. Does this mean that hypercalciuria is perhaps not so important for the development of a nephrocalcinosis or calcium urolithiasis as other factors in damage of tubule cells?

References

1. Adler S (1985) Urinary excretion of citrate – influence of metabolism and acid base conditions. In: Schwille PO et al. (eds) Urolithiasis and related clinical research. Plenum Press, New York London, p 173
2. Bell AJ, Lindner A (1984) Effects of verapamil and nifedipine on renal function and hemodynamics in the dog. Renal Physiol 7:329
3. Bichler KH, Kirchner C, Weiser H, Korn S, Strohmaier WL, Schmitz-Moormann P, Hanck A, Nelde HJ (1983) Influence of vitamin A deficiency on the excretion of uromucoid and other substances in the urine of rats. Clin Nephrol 20:32
4. Bichler KH, Strohmaier WL, Schanz F, Nelde HJ, Gaiser I, Schulze E, Schreiber M (1985) Zur Wirkung von Kalziumantagonisten (Nifedipin) auf die Nephrokalzinose und Kalziumausscheidung der Ratte. Urol Int 40:13
5. Chaignon M, Bellet M, Lucsko M, Rapoud C, Guedon J (1986) Acute and chronic effects of new calcium inhibition, nicardipine on renal haemodynamics in hypertension. Journées de L'Hypertension Artérielle
6. Dirks JH, Quamme GA, Sutton RAL (1985) Tubular handling of calcium. In: Fleisch H et al. (eds) Urolithiasis and related clinical research. Plenum Press, New York London, p 173
7. Füllgraf G, Meiforth A (1973) Calcium antagonist – effect of compound D 600 on renal tubulus resorption. Arch Pharmacol 276:243
8. Jasmin G, Solmoos B, Proschek I (1979) Therapeutic trials in hamster dystrophy. Annals New York Academy of Sciences Part V:338
9. Kinne-Saffran E, Kinne R (1974) Localization of a calcium-stimulated ATPase in the basallateral plasma membranes of the proximal tubule of rat kidney cortex. J Membrane Biol 17:263

10. Lang F, Deetjen P (1986) Persönliche Mitteilung
11. Lederballe PO, Mikkelsen E, Christensen NJ, Kornerup HJ, Pederson EB (1979) Effect of nifedipine on plasma renin, aldosterone and catecholamines in arterial hypertension. Eur J Clin Pharmacol 15:235
12. Leonetti G, Cuspidi C, Scamieri L, Terzoli L, Zanchetti A (1982) Comparison of cardiovascular, renal and humoral effects of acute administration of two calcium channel blockers in normotensive and hypertensive subjects. J Cardiovasc Pharmacol 4:319
13. Lossnitzer K, Janke J, Hein B, Stauch M, Fleckenstein A (1975) Disturbed myocardial calcium metabolism: a possible pathogenetic factor in the hereditary cardiomyopathy of the Syrian hamster. In: Fleckenstein A et al. (eds) Recent advances in studies on cardiac structures and metabolism, vol 6. Pathophysiology and morphology of myocardial cell alteration. University Park Press, Baltimore, p 207
14. Loutzenhiser R, Epstein M (1985) Effects of calcium antagonists on renal haemodynamics. Am J Physiol 249:F619
15. Marre M, Misumi J, Raemsch KD, Corvol P, Menard J (1985) Diuretic and natriuretic effects of nifedipine on isolated perfused rat kidneys. J Pharmacol Exp Therapeut 223:263
16. Massry SG, Kleemann CR (1972) Calcium and magnesium excretion during acute rise in glomerular filtration. J Lab Med 80:654
17. Nelde HJ, Bichler KH, Strohmaier WL, Schulze E, Gaiser I (1985) Histopathological changes in the kidney of the rat on an atherogenic diet. In: Schwille PO et al. (eds) Urolithiasis and related clinical research. Plenum Press, New York London, p 45
18. Ramm B, Hoffmann G (1976) Biomathematik und medizinische Statistik. Enke, Stuttgart
19. Strohmaier WL (1983) Vitamin A-Mangel und sein Einfluß auf das Verhalten von Uromucoid und anderer für die Harnsteingenese relevanter Parameter bei Ratten. Dissertation, Tübingen
20. Strohmaier WL, Bichler KH, Gaiser I, Schulze E, Nelde HJ, Schreiber M (1985) Urine and serum biochemistry relative to the risk of lithogenesis in rats on an atherogenous diet. In: Schwille PO et al. (eds) Urolithiasis and related clinial research. Plenum Press, New York, p 941
21. Suki WN (1976) Renal handling of calcium. Contr Nephrol 23:1
22. Ullrich KJ, Rumprich G, Klöss S (1976) Active Ca^2 reabsorption in the proximal tubule of the rat kidney. Pflügers Arch 364:223
23. Wallia R, Greenberg A, Puschett JB (1985) Renal hernodynamic and tubular transport effects of nitrendipine. J Lab Clin Med 105:498
24. Wrogemann K, Pena SDJ (1976) Mitochondrial calcium overload: a general mechanism for cellnecrosis in muscle diseases. The Lancet March 27:672
25. Yamaguchi I, Ikezawa K, Takada T, Kiyomoto A (1983) Studies on a new 1,5-benzothiazepine derivative (CRD-401). VI. Effects of renal blood flow and renal function. Jpn J Pharmacol 24:511
26. Zanchetti A, Leonetti G (1985) Natriuretic effect of calcium antagonists. J Card Pharmacol 7 (suppl 4):33

The Unique Chemical Composition of Nephrocalcinosis in Experimental Renal Insufficiency, Disturbances of Cellular Calcium Metabolism, and Protective Effect of Verapamil Against Nephrocalcinosis

M. S. Goligorsky[1], C. Chaimovitz[2], J. Rapoport[2], A. Kiryati[2], S. Lach[2], R. Kol[2], and J. Yehuda[2]

Clinical consequences of the deranged calcium metabolism are multiple. Mostly they manifest themselves dramatically as do renal stones or calcification of the soft tissues. These entities were known to ancient physicians and the Hippocrates' recommendation to avoid large quantities of milk for an infant's nursing (some should be substituted with a diluted wine) represents an indication of a deep intuitive awareness of what Sir Hamphry Davy will refer to in 1808 as calcium, and of its role in the "milk-alkali syndrome."

Ectopic Calcification in Uremia

The existence of widespread visceral calcification in chronic uremia has been known for at least 3 decades [1]. It was previously believed that hydroxyapatite was the main constituent of these calcificates. Contiguglia et al. [2] showed that in uremia diffraction patterns of heart, lung and skeletal muscle, as well as their Ca:Mg:P molar ratios resembled whitlockite. These investigators concluded that two types of calcificates exist in chronic uremia: non-visceral and arterial, containing hydroxyapatite; and visceral, containing whitlockite [2, 3].

Recent data revealing trace element content abnormalities in different tissues of uremic patients [4–9] indicate the need to accurately evaluate the chemical content of individual calcified fields. Moreover, in uremia little is known of the nature of renal calcification, while its contribution to progressive renal damage in chronic renal insufficiency is strongly suggested [10, 11].

Unique Chemical Composition of the Kidney in Uremia

We have previously demonstrated that calcium content of the kidneys of rats with Heyman's nephritis [12] or of rats subjected to 5/6 nephrectomy [13] is elevated. We pursued with these studies, utilizing subtotally nephrectomized rats as a convenient model of chronic renal insufficiency. The chemical composition of su-

[1] Jewish Hospital of St. Louis at Washington University School of Medicine, 216 South Kingshighway, St. Louis, Missouri 63110, USA, and [2] Ben-Gurion University Medical School, Beer-Sheba, Israel.

Nephrocalcinosis, Calcium Antagonists, and Kidney
Ed. by K.-H. Bichler and W.L. Strohmaier
© Springer-Verlag Berlin Heidelberg 1988

pracellular calcified deposits was examined by means of x-ray microanalysis and scanning electron microscopy, aided by x-ray diffraction and x-ray fluorometric analysis of the renal tissue [33].

Methods

Experiments were carried out on Charles River male rats weighing 200–250 g. Renal failure was induced by 5/6 nephrectomy using a flank incision. Sham-operated animals underwent flank incision and kidney decapsulation. Three weeks following operation, the animals were anesthetized with pentobarbital intraperitoneally and a blood sample was obtained from a tail artery and analyzed for creatinine, BUN, calcium and phosphorus. Kidneys were removed with special attention paid to removing the scarred capsule and to performing determinations on non-necrotic tissues. Light microscopy of hematoxylin-eosin and von Kossa-stained kidney slices obtained after these procedures proved the absence of coarse tissue damage due to the surgical procedure.

Sample preparation for scanning electron microscopy and x-ray microanalysis was performed with standard technique for light microscopy with paraffin-embedded sections. Such a preparation allowed subsequent light microscopy with x-ray microanalysis on SEM Jeol JSM with EDAX model 711.

X-ray diffraction analysis of kidney remnants was performed with an x-ray diffractometer PW 1050 (Philips) after incineration of the tissue at 500° C. The same samples were analyzed by means of x-ray fluorometry (PW 1410; Philips), under vacuum, using Rh-tube and coarse collimator. Samples containing 4.75 g $CaHPO_4$, and 0.05 g each SiO_2, Al_2O_3, MgO, TiO_2, and Fe_2O_3 were used as standard.

Uremic Model and Reliability of Nephrocalcinosis

Three weeks following 5/6 nephrectomy both BUN and creatinine plasma levels were elevated two-fold as compared to control rats. Calcium and inorganic phosphate plasma concentrations remained in the normal range, as did the $Ca \times P$ product. Von Kossa staining of renal sections and renal calcium content determination confirmed the consistent presence of nephrocalcinosis in the course of mild renal failure. A patchy distribution of delicate punctate von Kossa-positive material was found in the peritubular area (Fig. 1 A). This is in accordance with the observations in human renal failure of equivalent degree [14].

Scanning Electron Microscopy and X-Ray Microanalysis

Scanning electron microscopy of slices of renal tissue, adjacent to those examined by light microscopy, allowed visualization of similar particles under greater magnifications (Figs. 1 B, C). Concomitant x-ray microanalysis of individual particles revealed a broad elemental spectrum, as well as a variable calcium content of each deposit (Fig. 1 D). In accordance with previous data [2], deposits contained cal-

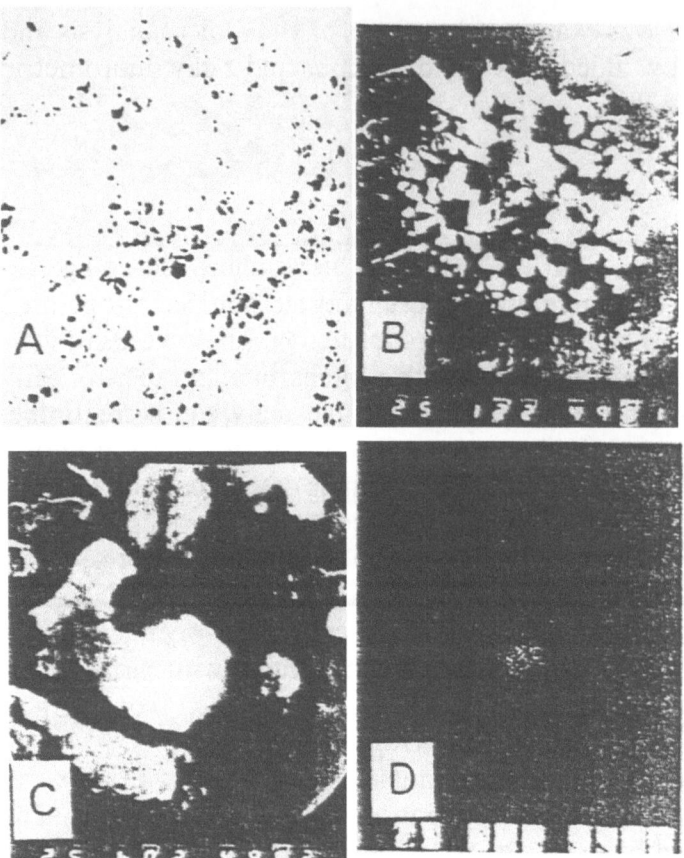

Fig. 1 A–D. Demonstration of calcificates in renal tissue of uremia rats: **A** von Kossa staining. × 800. **B** Scanning electron microscopy of calcificates × 1000, **C** and × 6000. **D** and secondary electron image of the same particle as in (C) demonstrating a high calcium content of the deposit (**D**). (From [35])

cium and magnesium. Interestingly, substantial amounts of aluminum and silicon and trace amounts of iron were constantly seen (Fig. 2). At the same time it was impossible to demonstrate any obvious calcium deposition in the samples obtained from the kidneys of control animals.

X-Ray Fluorescence Study

To ensure that these data on elemental content of calcified particles are not artifacts, spectroscopic examination of incinerated renal parenchyma was performed. Figure 3 demonstrates that the examined material contains the above-mentioned elements. However, the relative amount of aluminum, silicon, magnesium and iron obtained from the whole tissue analysis are much less than that found within an individual particle. Thus, it seems probable that these elements are preferentially concentrated within the calcified deposits and that their presence is not due to artifact.

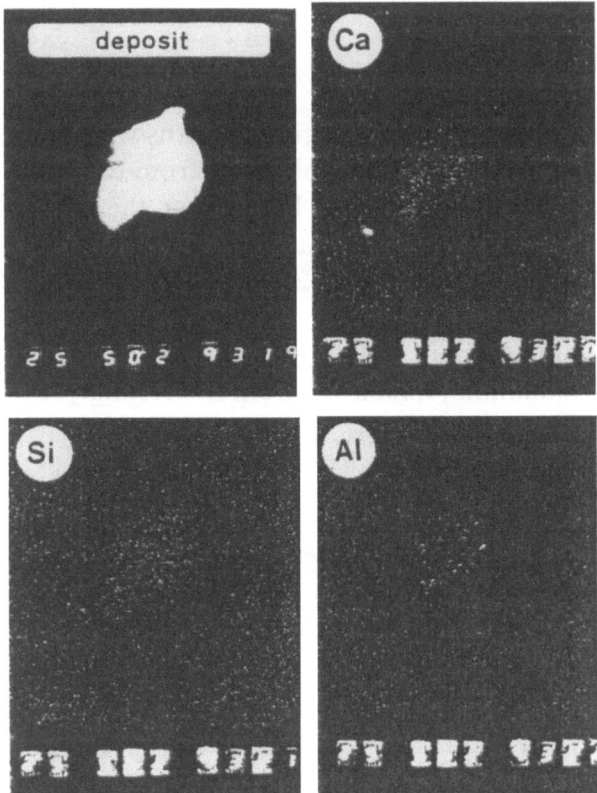

Fig. 2. Elemental mapping of a *calcified deposit*, containing Si and Al. (From [35])

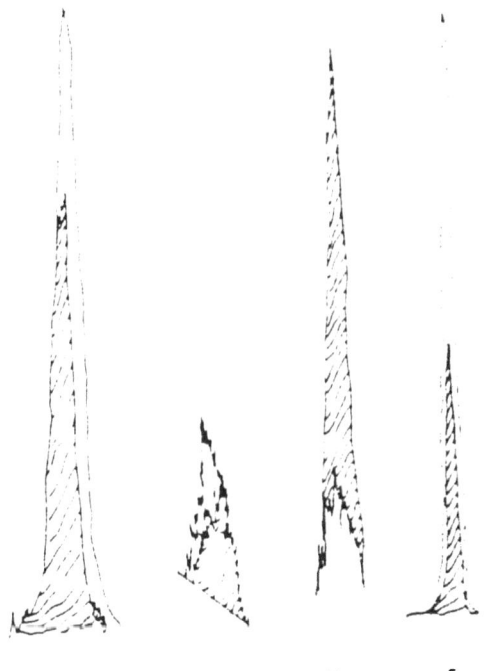

Fig. 3. X-ray fluorescence study of renal tissue of uremic rats (*open areas*) vs standard (*hatched areas*). (From [35])

Diffraction Patterns of Nephrocalcinosis

To further characterize uremic nephrocalcinosis, diffraction patterns were studied. Figure 4a–c depicts major types of diffractograms obtained from the incinerated kidney remnants. They differed from whitlockite. In contrast, the diffraction patterns of myocardial tissue of uremic rats, incinerated at 500° C (Fig. 4d), closely resemble those obtained by Contiguglia et al. [2] and characterized as whitlockite.

Our results show that, in addition to calcium and magnesium, the calcificates also contain significant amounts of aluminum, silicon and iron. Interestingly, cal-

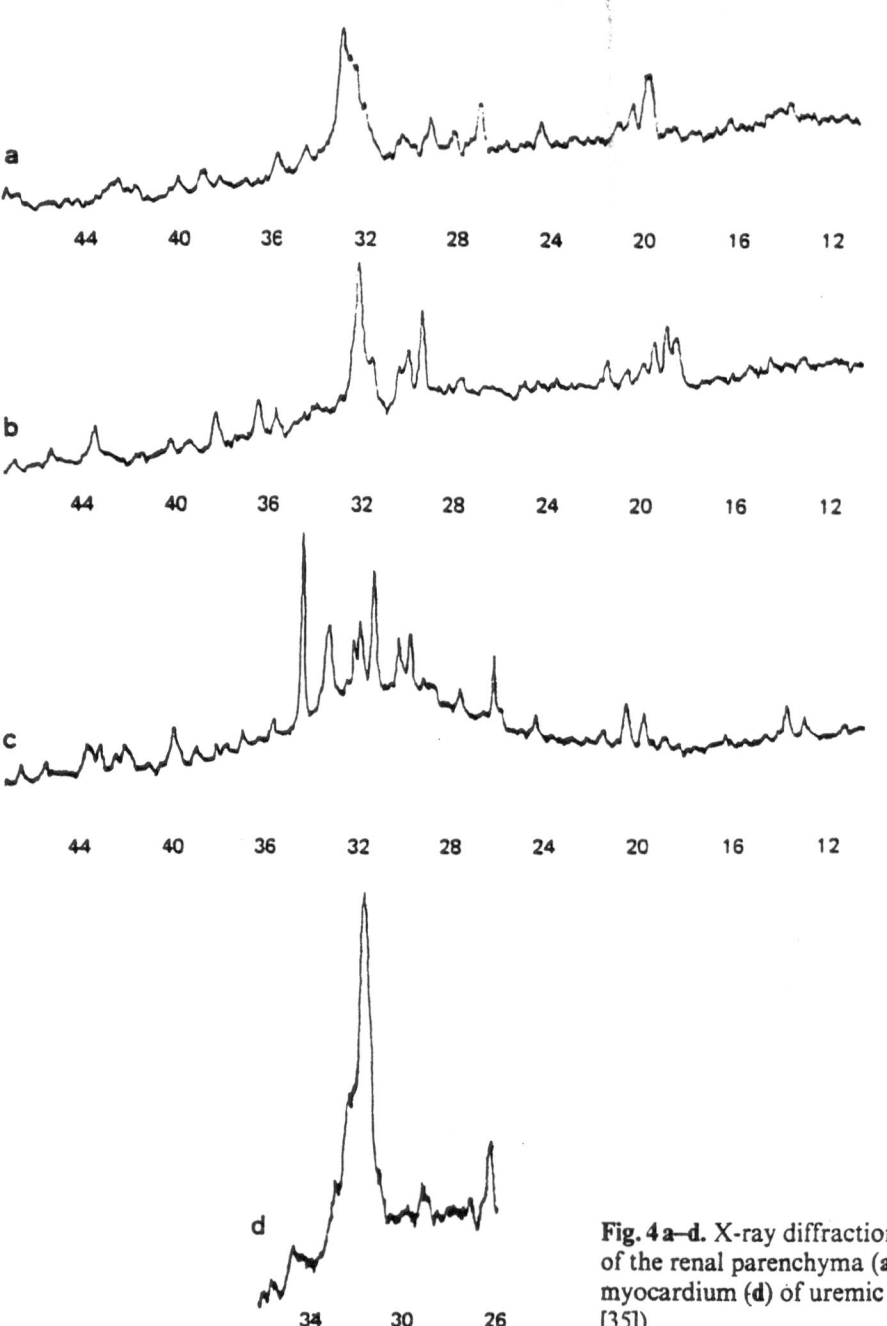

Fig. 4a–d. X-ray diffraction patterns of the renal parenchyma (a-c) and myocardium (d) of uremic rats. (From [35])

cificates were not of uniform composition, but contained the above elements in variable ratios. The inverse relationship between calcium and aluminum, and calcium and silicon content of calcificates was demonstrated. The significance of this inverse correlation is unclear.

Since the first warning about the potential toxicity of aluminum in the early '70s [5], accumulation of aluminum has been recognized as important clinically, in the development of osteomalacia and dialysis encephalopathy [4–8, 15]. Thus, the finding of aluminum deposition in early stages of uremia is of interest and could have therapeutic implications.

Our results confirm previous studies in human uremic subjects showing that the calcified deposits are heterogeneous in structure [14]. The reason for this lack of uniformity remains unclear. Possible causes are: (a) different origin of calcificates; (b) different environmental conditions along the nephron; (c) different stages in the calcification process. Next series of experiments was designed to address these possibilities.

Methods

Four weeks after surgery, rats were sacrificed and the kidney remnants were removed immediately with careful excision of scarred tissue. The following studies were performed on the renal cortex: (1) hematoxylin-eosin staining to ensure the completeness of the incision of the scarred tissue; (2) von Kossa staining to verify the presence of nephrocalcinosis in SNX rats and its absence in the control animals, and (3) preparation of thin sections (stained and unstained) for electron microscopy and x-ray microanalysis. Briefly, renal cortical samples were fixed with 2% glutaraldehyde in 0.1 M cocadylate buffer (pH 7.2) for 30 min. Samples were then rinsed twice in cocadylate buffer 0.1 M, pH 7.2 for 15 min and transferred to 1% osmium tetroxyde at 4° C for 1 h. After dehydration in graded ascending concentrations of ethanol, samples were embedded in Araldite 502. 0.8 and 2 μm sections were cut with an LKB-III ultramicrotome. 0.8 μm sections were stained with uranyl acetate and lead citrate. Stained sections were examined by means of transmission electron microscopy (Philips 201). Both stained and unstained 0.8 and 2 μm sections, mounted on copper grids, were examined with energy dispersive x-ray microanalysis (ED spectrum analyzer, Proxan, Elscint) combined with a Jeol 120 XC scanning-transmission electron microscope, operated at 80 kV acceleration voltage and a beam current of 80 μA. The measuring conditions were identical in all measurements. Zero adjustment was made with a copper Kα x-ray line. The x-ray spectra were collected over the range of 0–10 keV for 300 s. The selected area raster mode was used to collect spectra from an approximately 0.8 × 0.8 μm area at a magnification of × 30000. All x-ray spectra records were referred to background x-ray patterns with subsequent subtraction of the records. In the preliminary experiments, no significant difference in the sensitivity of detection of the elements of interest and their distribution in stained 0.8 μm and unstained 2 μm sections was found. In the stained sections the elemental spectrum additionally contained osmium, lead and uranium peaks. Because of the obvious resolutional advantages in studying stained sections, we shall refer below to the results obtained from thin stained sections.

In a separate series of experiments, the collagen fibers obtained from the rat tail were examined similarly by means of x-ray microanalysis.

Electron Microscopy

Figure 5 depicts proximal tubular cells in control (A) and SNX (B) animals. There is an obvious mitochondrial swelling in SNX rats, compared to control cells. These mitochondria appear to be loaded with an electron-dense material. In addition, the mitochondrial population in SNX rats shows evidence of disorganization, which includes condensed mitochondria, swollen cristae, herniation of internal membrane and the complete loss of internal structure. Frequency distribution analysis (with 500 objects examined) revealed that $62 \pm 7\%$ of the mitochondrial population were partially disorganized, while $27 \pm 7\%$ completely lost their internal structure.

The second abnormality was seen in the apical cell membrane (Fig. 6). The brush border of proximal tubular cells in SNX rats contained various poorly differentiated structures, resembling lamellar bodies (arrows).

An irregularity and loss of homogeneity of the tubular basement membrane of proximal epithelial cells in SNX rats represent the third major structural abnormality seen. Various inclusions within the tubular basement membrane, which were not observed in control animals, were a common finding in SNX rats [31].

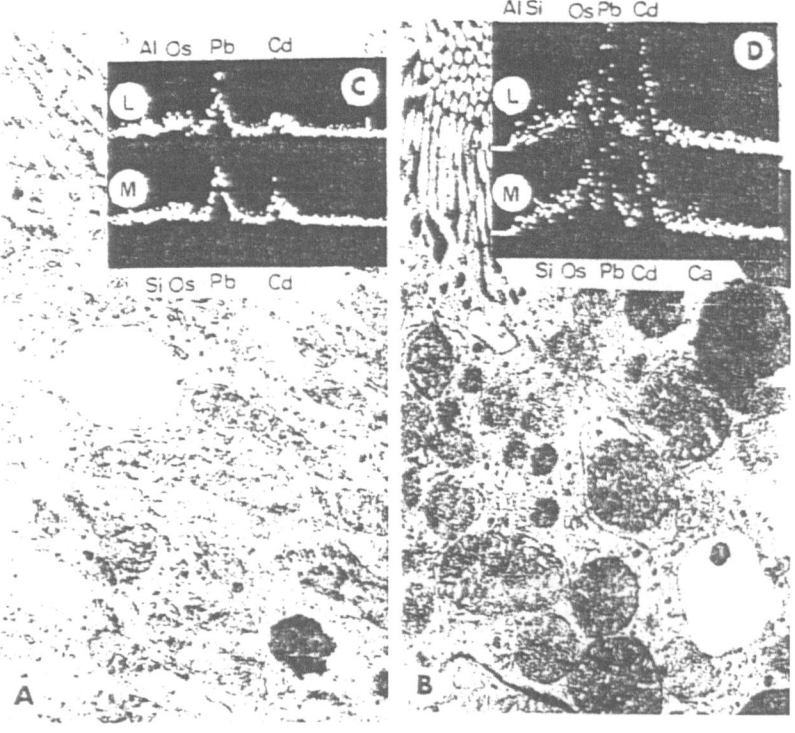

Fig. 5 A, B. Electron micrograph of proximal tubular cells from control (A) and SNX (B) rats. The tubular lumen is at the top. A bulk of the mitochondrial population in SNX rats consists of condensed forms, swollen mitochondria and mitochondria which have completely lost their internal structures. $\times 8850$. *C,D* – X-ray microanalysis of the corresponding preparations. *M* – mitochondria. *L* – lysosomes. (From [35])

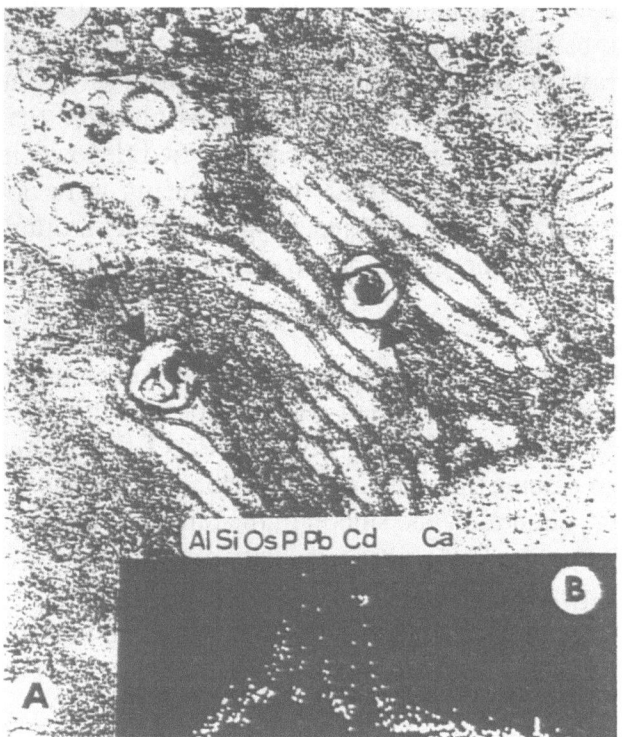

Fig. 6. A Electron micrograph of the brush border membrane of the proximal tubular epithelium of SNX rat with inclusions resembling lamellar bodies. **B** X-ray microanalysis of inclusions within the brush border membrane: peaks of aluminum, silicon and calcium can be seen. Peaks of osmium, lead and uranium are the result of staining procedures. (From [32])

X-Ray Microanalysis

Normal-looking mitochondria either in control, or in SNX animals, did not reveal any discernible patterns characteristic of calcium or aluminum. In contrast, the elemental content of disorganized mitochondria (Fig. 5 D) showed a peak characteristic of calcium, while secondary lysosomes revealed aluminum. A peak characteristic of silicon is present in both secondary lysosomes and mitochondria. It is remarkable that the tubular basement membrane in SNX rats contained calcium together with aluminum and silicon (Fig. 19). X-ray microanalysis of either the control tubular basement membrane, or background did not reveal these elements. Brush border membrane inclusions had elemental content similar to the tubular basement membrane, containing aluminum and silicon together with calcium (Fig. 6 B).

X-ray microanalysis of collagen did not reveal any discernible patterns characteristic of silicon or calcium.

A Possible Cellular Traffic of Ca, Al, and Si in the Pathogenesis of Uremic Nephrocalcinosis

We have shown that calcium and silicon appeared in disorganized mitochondria in proximal tubular cells of subtotally nephrectomized rats. It is noteworthy that the bulk of the mitochondrial population in SNX rats was represented by these disorganized forms. At the same time, aluminum and silicon patterns were obtainable from secondary lysosomes. Subcellular particles, resembling lamellar bodies, which crossed apical and basilar membranes, as well as the tubular basement membrane itself under x-ray microanalysis, displayed the entire set of the elements studied: calcium, aluminum, and silicon. It should be emphasized that these same elements were demonstrated within extracellular calcified deposits in uremic animals (Figs. 1 and 2). It is noteworthy that these elements were not found in control basement membranes and are absent in the background. Moreover, basically the same results have been obtained in unstained sections; the fact which makes unlikely the possibility of an artifact due to staining procedures.

In order to rule out a possible dietary source of the renal deposition of aluminum and silicon, metabolic studies have been carried out. It appeared that on a liberal diet, food consumption in the experimental group was lower than in the control. Hence, it is reasonable to assume that the dietary overload is not responsible for the deposition of calcium, aluminum and silicon in the renal parenchyma of uremic rats. This was further supported by a close similarity of the plasma aluminum concentrations in both experimental and control groups.

In a study of tissue composition in uremia [9] an almost two-fold increase in silicon content of the kidney was found. The possible sources of silicon in the kidney are as follows: (1) tubular fluid [22]; (2) fibrillar structures of connective tissue, rich in glycosaminoglycans and polyuronides [20]; and (3) mitochondrial granules [19].

Theoretically, all these sources could contribute to silicon accumulation within calcified deposits. The results of x-ray microanalysis of the collagen fibers showed that fibrillar structures most likely, were not an important source for silicon.

Concerning aluminum distribution within the cell, there is substantial evidence that this element is accumulated in lysosomes, and hence, can serve as a marker of these organelles. Following intraperitoneal aluminum administration to experimental animals, Galle [23] demonstrated aluminum in lysosomes using x-ray microanalysis. There is an extensive literature on aluminum retention in chronic renal failure [4–9]. Indeed, in our study aluminum was displayed in primary and secondary lysosomes and was absent in the mitochondria of uremic rats. Hence, in this setting aluminum within extracellular calcificates is of cellular origin, it most likely derives exclusively from the lysosomes.

Moreover, in these mildly uremic rats the plasma concentration of aluminum is kept within the normal range. Despite this, the filtration load of aluminum in the remaining nephrons is probably increased. This might serve as a primary source of aluminum accumulation in the lysosomes.

Calcium appears invariably within disorganized mitochondria in the proximal tubular epithelium of SNX rats. Calcium was further demonstrated in the tubular basement membrane and within brush border inclusions in the form of conjugates with aluminum and silicon. The fact that calcium and aluminum, which are separately sequestered within the different intracellular compartments, are found together on the polar surfaces of proximal tubular cells might suggest an interaction of the former compartments. In other words, lysosomal and mitochondrial interaction in the proximal tubular cells of uremic rats may contribute to the deposi-

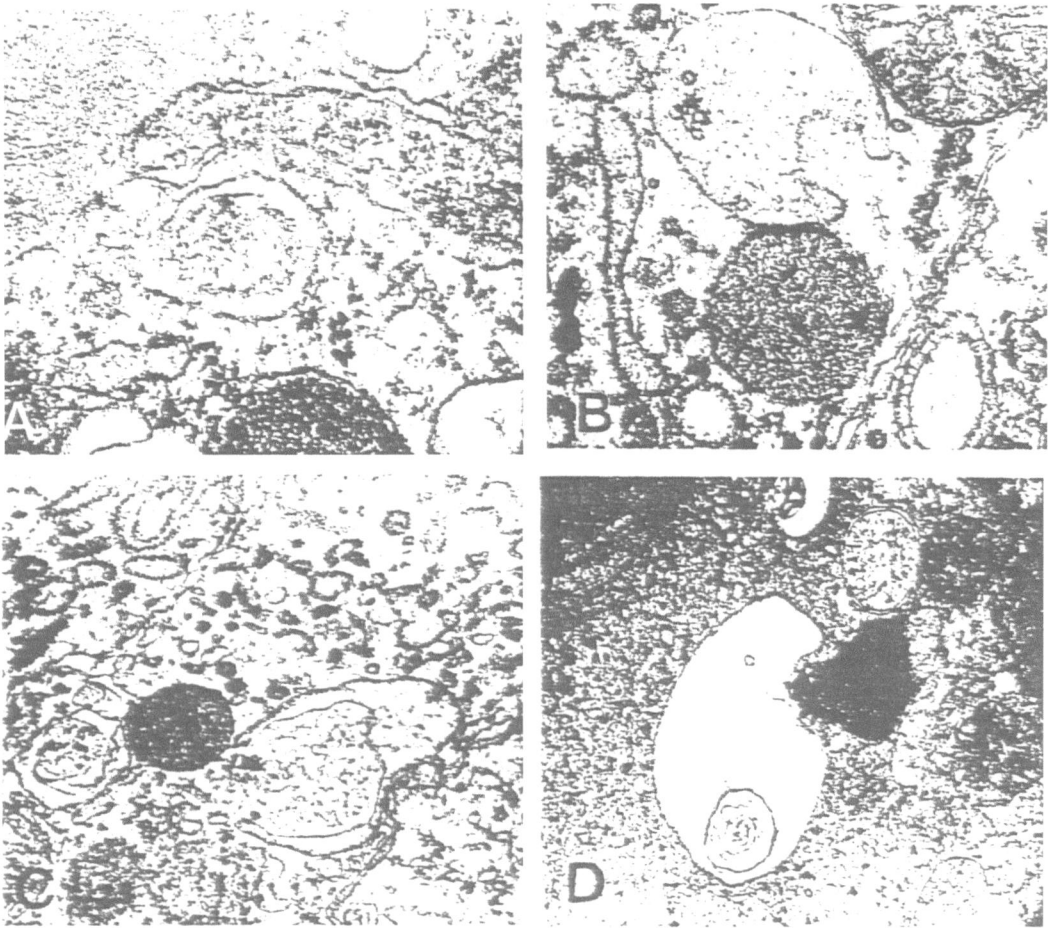

Fig. 7. Various forms of mitochondrial-lysosomal interaction in SNX rats (A–D)

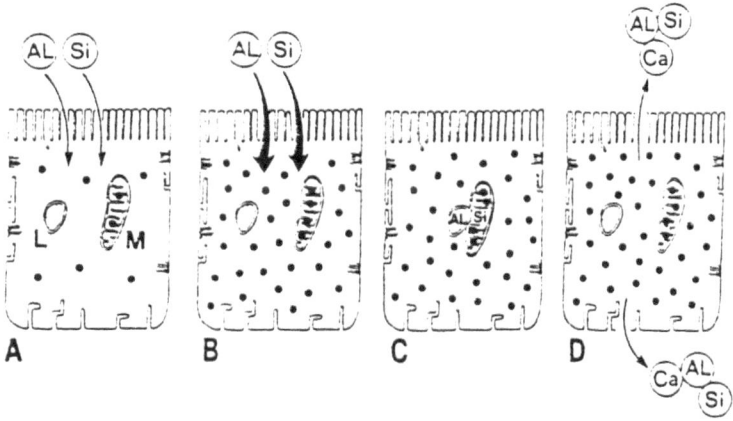

Fig. 8 A–D. Hypothetical sequence of the events leading to nephrocalcinosis in mild-to-moderate renal insufficiency. **A** Normal renal tubular cell (*black dots* represent Ca^{2+}). **B** Increased filtered load of *AL* and *Si* in hyperfiltering nephrons of the kidney remnant even before the concentration of these elements in plasma increases: deranged cell calcium metabolism. **C** Accumulation of *AL* and *Si* in lysosomes and in mitochondria: lysosomal-mitochondrial interaction. **D** Extrusion of the conglomerates of *Ca-AL-Si* towards the luminal or the basolateral surface

tion of calcium, aluminum and silicon in the renal parenchyma, resulting in nephrocalcinosis. Morphologic evidence for this conclusion is presented in Fig. 7. Subsequent stages of the possible interaction of these organelles in the pathogenesis of nephrocalcinosis are illustrated in Fig. 8. It should be emphasized that mitochondrial-lysosomal interaction is supposed to be of physiological significance, playing a role in the renovation of the cellular interior [24]. The reason why this process in experimental renal insufficiency leads to nephrocalcinosis remains to be established. In this respect, the low solubility of calcium alumino-silicates might play an important role.

Pathophysiology of Cellular Calcium Metabolism in Uremic Kidneys

Quibus in urina arenosa subsident, illis vesica calculo laborat (Hippocrate's Aphorismi, Sect. IV, 79). Since then, our understanding of the pathophysiology of the ectopic calcification, and of nephrocalcinosis in particular, was not enlightened dramatically. Since parathyroidectomy almost completely prevents the development of nephrocalcinosis in renal tissue, it has been suggested that secondary hyperparathyroidism may play an important role in its pathogenesis [10, 25].

The effect of parathyroid hormone (PTH) was studied by Borle [26] and Borle and Uchikawa [27], who demonstrated that the addition of PTH to the incubation medium led to an almost threefold increase in ^{45}Ca incorporation into renal cortical cells. This enhanced calcium incorporation was attenuated significantly following exclusion of phosphate and magnesium from the incubation medium; however, calcium incorporation still remained elevated. In chronic experiments on rats with the phosphate-induced secondary hyperparathyroidism, Borle and Clark [25] demonstrated a marked stimulation of renal cell calcium metabolism. It was shown that secondary hyperparathyroidism caused an intracellular accumulation of calcium, consequently leading to nephrocalcinosis. Our own data on calcium metabolism in SNX rats [31] are presented below.

Calcium Metabolism: Kidney Cortical Slices

In Fig. 9 A, renal cortical calcium content in SNX rats is depicted. The average calcium content in SNX rats was 32.3 ± 5.8 mmoles/kg dry weight, a value considerably higher ($p < 0.001$) than 14.0 ± 2.6 mmoles/kg dry weight in sham-operated animals.

^{45}Ca incorporation into kidney cortical slices from SNX rats (Fig. 9 B) revealed a 35% increase ($p < 0.05$) as compared to control values. La-resistant fraction of radiocalcium uptake exhibited even more profound differences between control and SNX animals. ^{45}Ca incorporation into La-treated renal cortical slices from SNX rats revealed a 50% increase ($p < 0.001$), compared to similarly treated slices in the sham group. This data indicates that the increase in ^{45}Ca incorporation into slices from SNX rats is due to intracellular accumulation of radiocalcium. ^{45}Ca washout curves obtained from the SNX group demonstrated that the radioactivity of slices remain elevated ($p < 0.05$) as compared to the sham group, up to 30 min of incubation (Fig. 9 C).

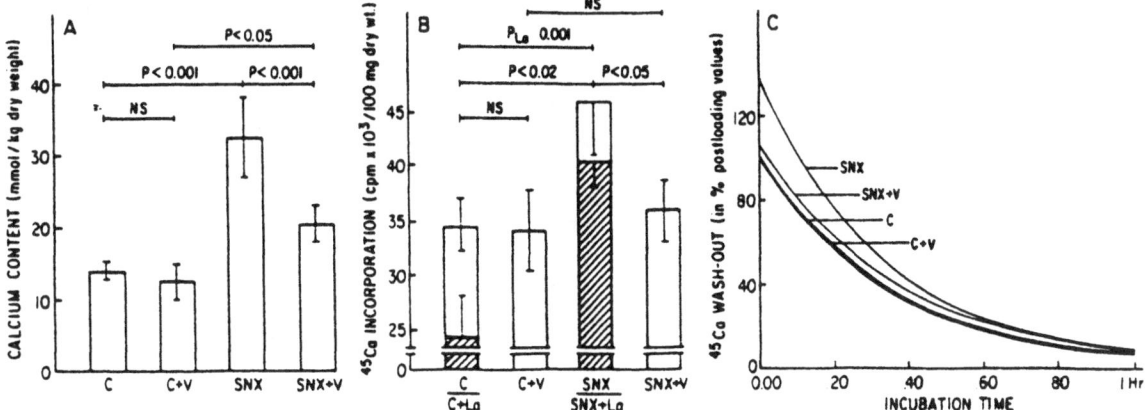

Fig. 9. A Calcium content of the renal cortex in experimental groups: *C*, control; *C* + *V*, control rats chronically treated with verapamil; *SNX*, subtotally nephrectomized rats; *SNX* + *V*, subtotally nephrectomized rats treated chronically with verapamil. Vertical bars represent mean ± SD. **B** ^{45}Ca incorporation into renal cortical slices in four experimental groups. *Dotted bars* represent La-resistant fraction of ^{45}Ca uptake. **C** ^{45}Ca washout curves in four experimental groups (in percent to control postloading values). Every curve is constructed from five time points of 2, 5, 10, 30, and 60 min (five slices for each point) by least square fitting. [31]

A Possible Role of Parathyroid Hormone in the Disturbances of Cellular Calcium Metabolism and in the Development of Nephrocalcinosis

Secondary hyperparathyroidism in rats fed with a high phosphate diet was associated with renal cellular calcium accumulation within 24 h [25]. Accumulation of calcium within the mitochondrial compartment was prominent in this model of nephrocalcinosis. Similarly, repeated injections of PTH in mice resulted in nephrocalcinosis. Mitochondrial disorganization was detected in the early stage of the process even before the appearance of nephrocalcinosis [28, 29]. It is of interest that in nephrocalcinosis produced by vitamin D overdose, mitochondria are also the primary site of the cellular damage [30]. When 2,4-dinitrophenol was inoculated together with vitamin D, the dosage of the latter required to induce nephrocalcinosis was reduced significantly. It was thus postulated by Scarpelli [30] that mitochondria exert a protective effect against nephrocalcinosis and their obliteration accelerates the development of renal calcification.

Recently, Drs. D. Loftus, K. Hruska and one of us, examined a possible role of PTH in cellular calcium metabolism using microspectrofluorometry of fura-2-loaded cultured proximal tubular cells [34]. Cells grown on glass coverslips were loaded with 20 μ*M* fura-2/AM for 60 min and then mounted into a Sykes-Moore chamber. The chamber was placed on the stage of a Nikon Optiphot microscope equipped for epifluorescence. The Sykes-Moore chamber was perfused with oxygenated Krebs-Henseleit-bicarbonate solution, pH 7.4, at a rate of 3–4 ml/min for about two hours. Following this washing procedure, the cytoplasm of the cells exhibited homogeneous fluorescence of fura-2.

Focusing on a single cell was performed using UV-40 oil immersion objective and a pinhole diaphragm (Fig. 10) and corrected via a CF-eyepiece lens. Cells were excited alternatively with UV light at 340 nm and 380 nm wavelengths.

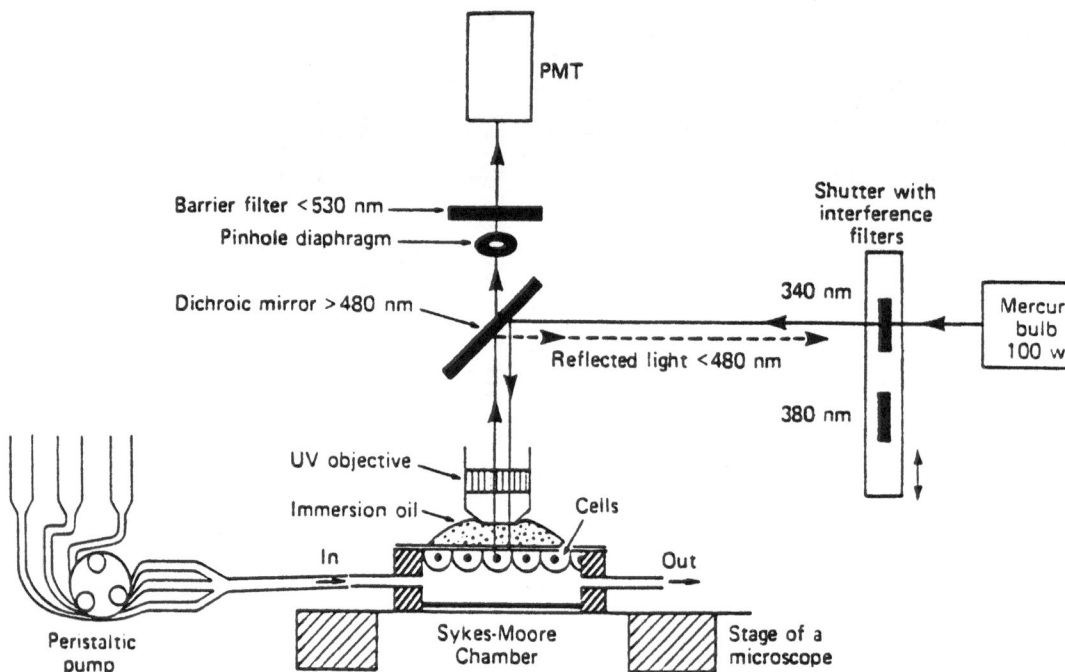

Fig. 10. Experimental set-up for monitoring of cytoplasmic calcium concentration in a single cell. (From [34])

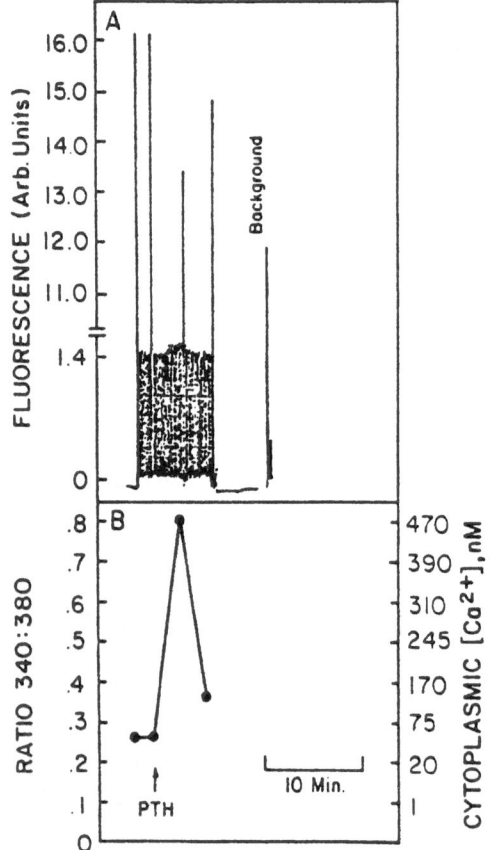

Fig. 11. Effect of PTH on cytoplasmic calcium concentration. (From [34])

Emitted light was collected at 480–530 nm. The conversion of 340/380 ratios to the respective concentrations of cytoplasmic Ca^{2+} was performed using the calibration curve.

$10^{-8}M$ PTH in the perfusate caused a rapid, reversible increase in cytoplasmic Ca^{2+} concentration (Fig. 11). The whole cycle of Ca^{2+} transient with a recovery of cytoplasmic Ca^{2+} concentration was completed within 5 min. Repeated pulses of PTH with an interval of 20 min did not affect neither the amplitude nor the resting state, although the concentration of cyclic AMP in the effluent progressively declined with every sequential pulse of PTH (Fig. 12). When the frequency of PTH pulses was increased (every 5 min), the amplitude of Ca^{2+} transients declined and the recovery of cytoplasmic Ca^{2+} was incomplete (Fig. 13). This phenomenon occurring acutely during excessive stimulation of the proximal tubular cells with PTH may represent a prototype of deranged cellular metabolism of calcium in hyperparathyroidism. The dissociation between cyclic AMP and cy-

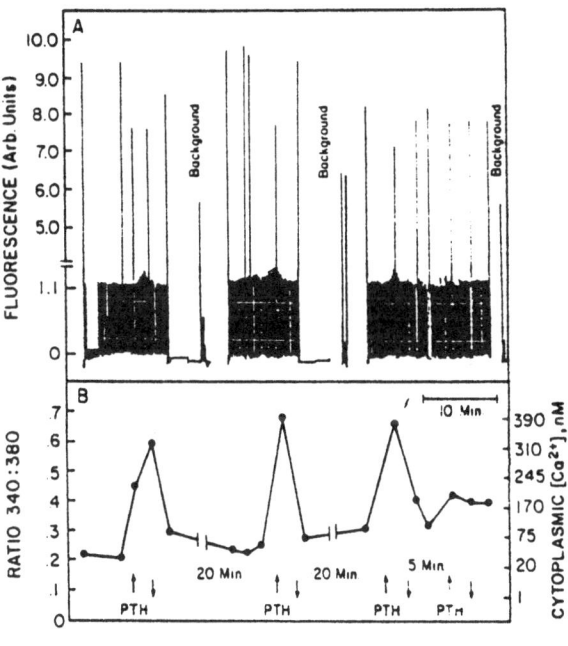

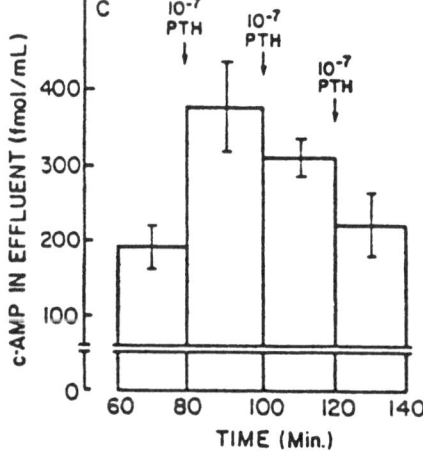

Fig. 12 A–C. Sequential perfusion of proximal tubular cells with PTH (From [34]). **A, B** Cytoplasmic calcium transients. C concentration of cyclic AMP in the effluent

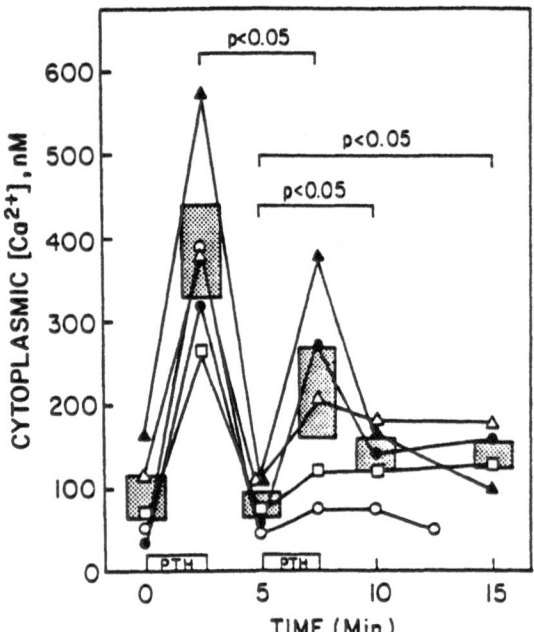

Fig. 13. The effect of frequent PTH pulses on cytoplasmic calcium concentration. (From [34])

toplasmic Ca^{2+} responsiveness to PTH could reflect an intermediate step in renal pathophysiology of hyperparathyroidism.

Protective Effect of Verapamil Against Cellular Calcium Disturbances and the Development of Nephrocalcinosis

Since intracellular calcium accumulation plays a pivotal role in the expression of PTH-induced nephrocalcinosis, it follows that attenuation of intracellular calcium influx would be expected to be of value in ameliorating this effect of PTH.

Recently, calcium channel blockers have been shown to inhibit various specific effects of PTH in various cells [35–40]. Furthermore, attenuation of calcium influx by verapamil has been demonstrated in the various epithelial layers as well as in the renal brush border membrane preparation [39, 40]. Thus, it might be suggested that the calcium channel blockers may prevent or attenuate uremic nephrocalcinosis.

A Direct Effect of Verapamil on Cytoplasmic Calcium Concentration

The issue of a direct effect of verapamil on renal tubular epithelium is controversial. Malis et al. [41] was not able to demonstrate a direct effect of verapamil on renal tubular cells. In several other laboratories specific binding of verapamil and its effect on renal tubular epithelium have been documented [42, 43]. The issue is further complicated by the existing uncertainty regarding molecular pharmacology of verapamil and by an emerging concept on its possible action on calmodulin [44].

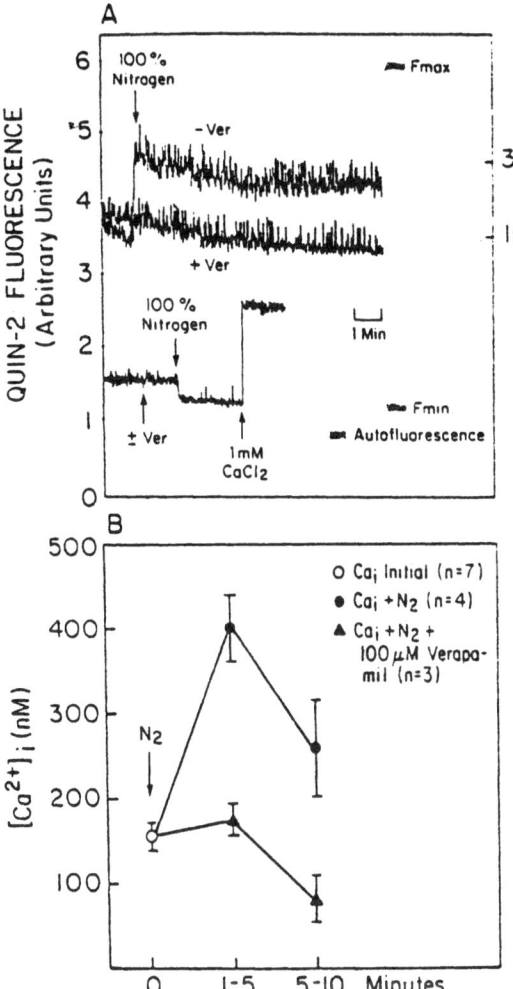

Fig. 14 A, B. The effect of anoxic insult on cytoplasmic calcium concentration. **A** Representative tracings of the nitrogen-induced changes of quin-2 fluorescence in proximal tubular cells. Verapamil pretreatment abolished calcium transient. The lower tracing represents changes in quin-2 fluorescence of cells incubated in a nominally calcium-free Krebs-Henseleit buffer. Initial cytoplasmic calcium concentration of 45 nM exhibited a further decrease upon exposure to nitrogen. Verapamil did not modify this effect. **B** Kinetics of calcium transients in verapamil-treated and untreated proximal tubular cells following anoxic insult (summary of the results obtained in 7 separate experiments). (From [45])

In view of these difficulties, we attempted to examine a direct end effect of verapamil on cytoplasmic-free calcium concentration in proximal tubular cells exposed to different Ca^{2+} mobilizing stimuli [45]. Anoxic insult (Fig. 14) as well as hormonal stimulation (PTH and α_1-adrenergic agonists) (Fig. 15) resulted in an increase in cytoplasmic calcium concentration. This calcium transient was prevented by verapamil. It is possible that verapamil interferes with α_1-adrenergic stimulation at the level of the receptor-operated Ca^{2+} channel; the phenomenon that has been demonstrated in platelets, renal cortex, heart muscle and in hepatocytes [42, 46–48]. The molecular mechanism of verapamil action during anoxic insult or PTH administration (either membrane-mediated or calmodulin-dependent) remains unknown. Nevertheless, these results demonstrated that verapamil interrupted the chain of events leading to an increase in cytoplasmic calcium concentration during various provocative stimuli employed. It is possible that an analogous mode of action could underlie the effect of this compound during its chronic administration in SNX rats (Fig. 16).

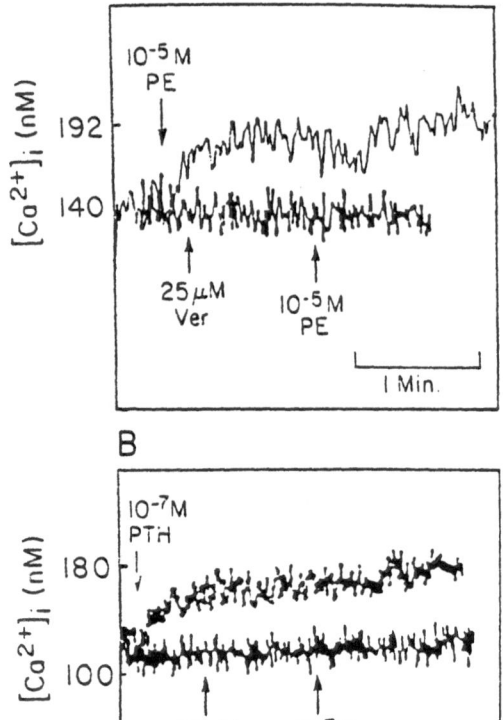

Fig. 15 A, B. Monitoring of quin-2 fluorescence in verapamil-treated and untreated proximal tubular cells exposed to Ca^{2+}-mobilizing hormones. The effect of α_1-adrenergic agonist phenylephrine (A), and PTH (B) on cytoplasmic calcium concentration is prevented by verapamil pretreatment. (From [45])

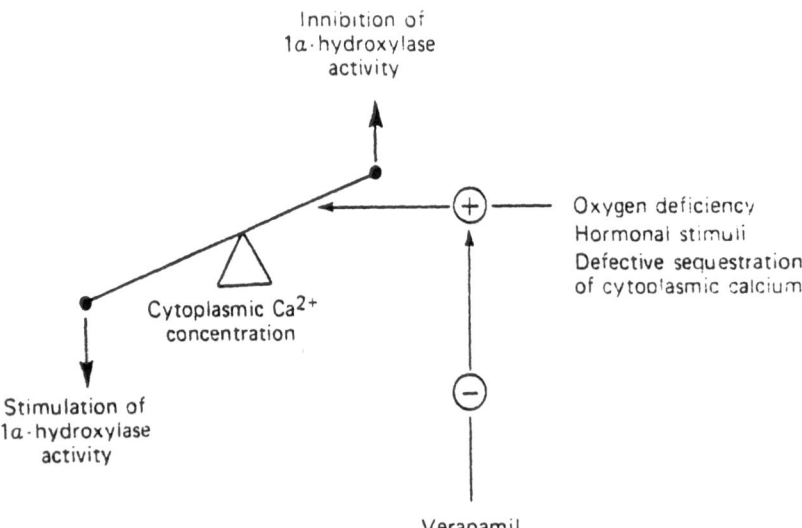

Fig. 16. Hypothetical mode of action of verapamil on $1,25(OH)_2D_3$ production in SNX rats. Various stimuli (e.g. oxygen deficiency, catecholamines, PTH, insufficent calcium sequestration, etc.) could lead to an inhibition of 1a-hydroxylase activity by increasing (+) cytoplasmic calcium concentration. Verapamil may prevent this inhibition by abolishing (−) the effect of the above stimuli on cytoplasmic calcium concentration. [45]

The Effect of Verapamil on Cellular Calcium Metabolism in SNX Rats

We next examined the renal calcium content, morphologic criteria of nephrocalcinosis and calcium kinetics in mild chronic renal insufficiency during verapamil treatment [31]. Obviously, there is not much sense to study the effects of verapamil in severe renal failure (7/8 nephrectomy); therefore the experiments were performed in rats with mild-to-moderate renal insufficiency. Von Kossa stained sections of kidney remnants obtained from SNX rats showed a delicate punctate calcification of renal tubular cells in the region of the tubular basement membrane (Fig. 17 A). In contrast, calcification was demonstrated to an obviously lesser extent in sections of verapamil-treated SNX animals (Fig. 17 B).

The data on calcium metabolism in renal cortical slices were presented above (Fig. 9). In verapamil treated SNX rats, renal cortical calcium content $(21.3 \pm 4.4$ mmoles/kg dry weight) was significantly lower ($p < 0.001$) than that seen in SNX rats, but still higher ($p < 0.001$) than control values.

^{45}Ca incorporation into renal cortical slices from SNX + V animals was similar to that obtained in control animals of both verapamil treated and untreated groups. Radiocalcium washout curves obtained from both control groups and verapamil treated SNX animals were identical (Fig. 9) as opposed to untreated SNX rats.

Verapamil and Uremic Nephrocalcinosis

Ultrastructurally, there is an obvious swelling and disorganization of mitochondria in uremic animals both treated and untreated. However, the prevalence of disorganized forms in the SNX group, as compared to SNX + V animals is prominent. Mitochondria of SNX rats revealed swelling of condensed forms, electron density, herniation of internal membrane, destruction of cristae and the complete loss of internal structure. These findings are in obvious contrast to the rel-

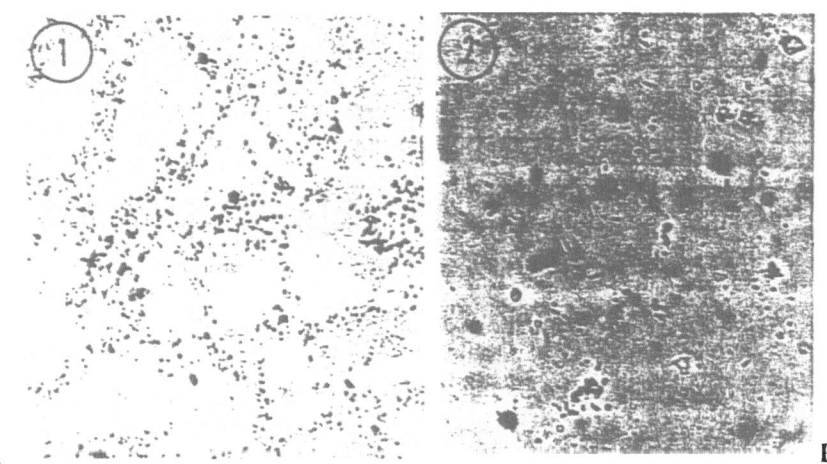

A   B

Fig. 17. A (1) Von Kossa staining of kidney section obtained from SNX rat (Magnification × 400). B (2) Von Kossa staining of kidney section from SNX rat treated with verapamil (Magnification × 400). (From [31])

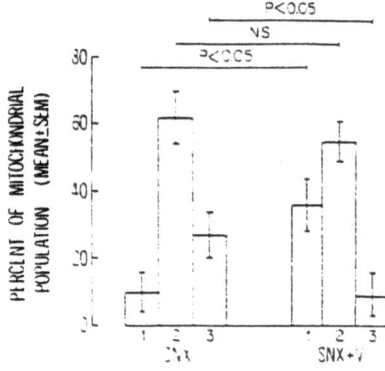

Fig. 18. Comparative semiquantitative analysis of mitochondrial damage in verapamil treated and untreated SNX animals. The numbers represent: *1.* percent of intact mitochondria; *2.* partially disorganized mitochondria; *3.* percent of mitochondria with completely distorted internal structure. (From [31])

atively intact state of mitochondrial structure in SNX rats treated with verapamil. The comparative analysis of mitochondrial ultrastructure in SNX and SNX+V animals is depicted in Fig. 18. Five hundred randomly selected objects were examined in each group of treated and untreated animals. Mitochondrial ultrastructure was arbitrarily subdivided as follows: (1) intact mitochondria; (2) partial disorganization (destruction of at least one-third of the crisate and/or herniation of internal membrane); (3) complete destruction of the internal structure. Verapamil treatment is associated with an increase ($p < 0.05$) in the fraction of intact mitochondria and a decrease ($p < 0.05$) in the subpopulation of completely distorted forms.

The tubular basement membrane represents another site of severe structural alteration in SNX rats. As depicted in Fig. 19. the tubular basement membrane

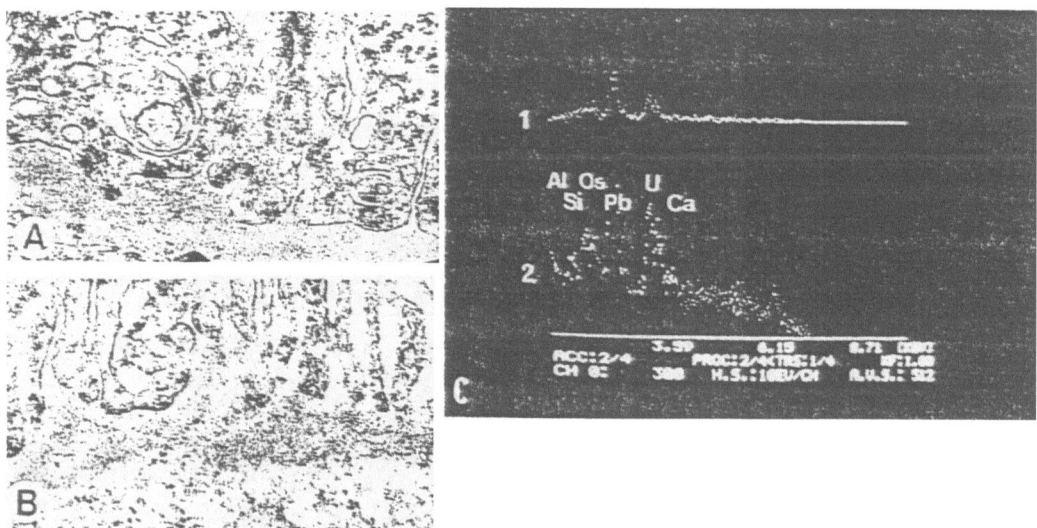

Fig. 19 A–C. Electron micrographs of the tubular basement membrane in SNX + V (**A**) and SNX (**B**) rats. Note the loss of homogeneity and integrity of the tubular basement membrane in the SNX rat, as compared to intact membrane in SNX + V animal (Magnification ×30000). **C** represents an x-ray microprobe analysis of the tubular basement membrane in SNX + V (*1*) and SNX (*2*) rats. The latter reveals calcium, phosphorus, aluminum, and silicon patterns characteristic of linear mineralization of the tubular basement membrane (*2*). This phenomenon is ameliorated in SNX + V rats (*1*). (From [31])

in this condition appears irregular and non-homogeneous. X-ray microanalysis of such a membrane (Fig. 19) shows the consistent patterns of calcium, phosphorus, aluminum and silicon, indicating that structural abnormalities are associated with the linear mineralization of the tubular basement membrane. Chronic verapamil treatment of SNX rats leads to a structural sparing of the tubular basement membrane (Fig. 19) and considerably reduced mineralization (Fig. 19).

One of the major factors controlling 1a-hydroxylase activity is the cytoplasmic calcium concentration; an increased concentration suppressed 1a-hydroxylase activity of mitochondrial preparations obtained from vitamin D deficient animals [49, 50]. Similarly, in vivo studies in rats receiving high doses of diphosphonates resulted in suppression of 1a-hydroxylase activity, a reduction in intestinal calcium transport and renal calcium accumulation [51-53].

We have demonstrated [45] that chronic verapamil administration to subtotally nephrectomized rats, while improving calcium metabolism, resulted in a twofold elevation of plasma levels of $1,25(OH)_2D_3$ and in an improvement of duodenal calcium absorption. It is tempting to speculate that these phenomena are sequentially connected. The primary effect of verapamil may be to prevent cytoplasmic calcium bursts in renal tubular cells of uremic animals. This in turn, may "release" the feedback regulation of 1a-hydroxylase activity and increase production of $1,25(OH)_2D_3$ as schematically depicted in Fig. 16. The final effect could be an improved duodenal calcium absorption.

Conclusions

In summary, studies of uremic nephrocalcinosis revealed that it develops early in the course of renal insufficiency. The unique conditions developing in the functioning nephrons after the reduction of renal mass, particularly an increased filtered load of Al and Si, together with the developing hyperparathyroidism, can hypothetically trigger the formation of calcium-alumosilicates with an irreversible deposition of these highly stable chemical structures in the renal parenchyma. Still, the deranged cellular calcium metabolism is essential for their formation. Chronic verapamil administration is associated with an improvement of cellular calcium metabolism. This is at least in part, a direct effect of verapamil upon renal tubular epithelium, which results in amelioration of nephrocalcinosis. Recent studies [54] indicated that verapamil administration alleviated the rate of progression of functional deterioration in subtotally nephrectomized animals.

Acknowledgement. The authors wish to express their gratitude to Mrs. Karon Hertlein for expert secretarial assistance.

These studies were supported in part by grant No. 39612014 from the Israeli Ministry of Health.

References

1. Mulligan RM (1974) Metastatic calcification. Arch Path 43:177–230
2. Contiguglia SR, Alfrey AC, Miller NL, Runnels DE, LeGeros RR (1973) Nature of soft tissue calcification in uremia. Kidney Intl 4:229–235
3. Parfitt AM (1976) Soft tissue calcification in uremia. Dial Transplant 5:17–24
4. Berlyne GM (1979) Aluminium toxicity in man. Mineral Electrolyte Metab 2:71–74
5. Berlyne GM, Ben-Ari J, Pest D, Wemberger J, Stern M, Gilmore GR, Levine R (1970) Hyperaluminemia from aluminium resin in renal failure. Lancet II:494–396
6. Smythe WR, Alfrey A, Caswell PW, Crouch CA, Ibels LS, Kubo H, Nuneeley LL, Rudolph H (1982) Trace element abnormalities in chronic uremia. Ann Intern Med 96:302–310
7. Sanstead HH (1980) Trace elements in uremia and hemodialysis. Am J Clin Nutr 33:1501–1508
8. Alfrey AC, Hegg A, Craswell P (1980) Metabolism and toxicity of aluminium in renal failure. Am J Clin Nutr 33:1509–1516
9. Indraprasit S, Alexander GV, Gonick HC (1974) Tissue composition of major and trace elements in uremia and hypertension. J Chron Dis 27:135–161
10. Ibels LS, Alfrey AC (1981) Effects of thyroparathyroidectomy, phosphate depletion and diphosphate therapy on acute uremic extraosseous calcification in the rat. Clin Sci 61:621–626
11. Ibels LS, Alfrey AC, Huffer WE, Craswell PW, Weil R (1981) Calcification in end-stage kidneys. Am J Med 71:33–37
12. Goligorsky M (1977) Divalent ions in Heyman's nephritis. Vrach Delo 10:100–105
13. Goligorsky M, Fedorov A (1978) Nephrocalcinosis in chronic renal failure. Urol Nephrol 6:32–36
14. Kuzela DC, Huffler WE, Conger JD, Winter SD, Hammond WS (1977) Soft tissue calcification in chronic renal dialysis patients. Am J Pathol 86:403–424
15. Hodsman AB, Sherrard PJ, Alfrey A, Ott S, Brickman AS, Miller NL, Maloney NA, Coburn JW (1982) Bone aluminium and histomorphometric features of renal osteodystrophy. J Clin Endo Metab 54:539–546
16. Bommer J, Ritz E, Waldherr R (1981) Silicone-induced splenomegaly. N Eur J Med 305:1077–1079
17. Leong ASY, Disney APS, Gove DW (1982) Spallation and migration of silicone from bloodpump tubing in patients on hemodialysis. N Eur J Med 306:135–140
18. Saldanha LE, Rosen VJ, Gonick HC (1975) Silicon nephropathy. Am J Med 59:95–103
19. Mehard CW, Volcani BE (1976) Silicon-containing granules of rat liver, kidney and spleen mitochondria. Electron probe microanalysis. Cell Tissue Res 166:255–267
20. Schwarz K (1973) A bound form of silicon in glycosaminoglycans and polyuronides. Proc Natl Acad Sci 70:1608–1612
21. Lehninger AL (1977) Mitochondria and biological mineralization process: an exploration. Horizons Biochem Biophys 4:1–30
22. Dobbie JW, Smith MJB, Abdullah ARAS (1981) Silicon and the kidney. Proc 8th Intl Congress Nephrol, Athens, pp 1030–1034
23. Galle P (1983) The role of lysosomes in the renal concentration of mineral elements. In: Hamburger, Grosnier, Grunfeld, Maxwell (eds) Advances in nephrology, vol 12. Year Book Medical Publishers, Chicago, pp 85–99
24. Kim K (1983) Pathological calcification. In: Pathobiology of cell membranes, vol 3. Academy Press, New York, pp 117–155
25. Borle AB, Clark I (1981) Effects of phosphate-induced hyperparathyroidism and parathyroidectomy on rat kidney calcium in vivo. Am J Physiol 241:E136–E141
26. Borle AB (1970) Kinetic analysis of calcium movements in cell cultures. III. Effects of calcium and parathyroid hormone in kidney cells. J Gen Physiol 55:163–186
27. Borle AB, Uchikawa T (1978) Effects of parathyroid hormone on the distribution and transport of calcium in cultured kidney cells. Endocrinol 102:1725–1732
28. Caulfield JB, Schrag PE (1964) Electron microscopic study of renal calcification. Am J Pathol 44:365–382
29. Berry JP (1970) Nephrocalcinose experimentale par injection de parathormone. Etude an microanalyseur a sonde electronique. Nephron 7:97–116

30. Scarpelli DG (1965) Experimental nephrocalcinosis. A biochemical and morphologic study. Lab Invest 14:123–141
31. Goligorsky MS, Chaimovitz C, Rapoport J, Goldstein J, Kol R (1985) Calcium metabolism in uremic nephrocalcinosis: preventive effect of verapamil. Kidney Intl 27:774–779
32. Goligorsky MS, Chaimovitz C, Nir Y, Rapoport J, Kol R, Yehnda J (1985) X-ray micro-analysis of uremic nephrocalcinosis: cellular distribution of calcium, aluminum and silicon. Mineral and Electrolyte Metabl 11:301–308
33. Goligorsky MS, Chaimovitz C, Rapoport J, Zevin L, Kiryati A, Lach S (1983) X-ray micro-analysis of uremic nephrocalcinosis. Nephron 35:89–93
34. Goligorsky MS, Loftus D, Hruska KA (1986) Microspectrofluorometric monitoring of cytoplasmic Ca^{2+} in individual proximal tubular cells in culture. Am J Physiol 251:F938–F944
35. Bogin E, Massry SG, Harary I (1981) Effect of parathyroid hormone on rat heart cells. J Clin Invest 67:1215–1227
36. Katon Y, Klein KL, Kaplan RA, Sanborn WG, Kurokawa K (1981) Parathyroid hormone has a positive inotropic action in the rat. Endocrinol 109:2252–2254
37. Bogin E, Massry SW, Levi J, Djaldeti M, Bristol G, Smith J (1982) Effect of parathyroid hormone on osmotic fragility of human erythrocytes. J Clin Invest 69:1017–1025
38. Wrobel J, Michalska L (1977) The effect of verapamil on intestinal calcium transport. Eur J Pharmacol 45:385–387
39. Humes HD, Simmons CF, Brenner BM (1980) Effect of verapamil on the hydroosmotic response to antidiuretic hormone in toad urinary bladder. Am J Physiol 239:F250–F257
40. Sommermeyer MG, Knauss TC, Weinberg JM, Humes HD (1983) Characterization of Ca^{2+} transport in rat renal brush-border membranes and its modulation by phosphatidic acid. Biochem J 214:37–46
41. Malis C, Cheung J, Leaf A, Bonventre J (1983) Effect of verapamil in models of ischemic acute renal failure in the rat. Am J Physiol 14:F735–F742
42. Motulsky HJ, Snavely MD, Hughes RJ, Insel PA (1983) Interaction of verapamil and other calcium channel blockers with α_1- and α_1-adrenergic receptors. Circ Res 52:226–231
43. Saggerson ED, Carpenter CA (1980) Effect of compound D-600 (methoxyverapamil) on gluconeogenesis and on acceleration of the process by α-adrenergic stimuli in rat kidney tubules. Biochem J 190:283–291
44. Janis RA, Triggle DJ (1983) New developments in Ca^{2+} channel antagonists. J Med Chem 26:775–785
45. Goligorsky MS, Chaimovitz C, Shany S, Rapoport J, Sharony Y, Haichenko J (1986) Verapamil improves defective duodenal calcium absorption in experimental chronic renal failure. Mineral and Electrolyte Metab 12:363–370
46. Blackmore PF, El-Refai M, Exton J (1979) Alpha-adrenergic blockade and inhibition of A23187-mediated Ca^{2+} uptake by the calcium antagonist verapamil in rat liver cells. Molec Pharmac 15:498–506
47. Barnathan ES, Addonizio VP, Shattil SJ (1982) Interaction of verapamil with human platelet α-adrenergic receptors. Am J Physiol 242:H19–H23
48. Nayler WG, Thompson JE, Jarrott B (1982) The interaction of calcium antagonists (slow channel blockers) with myocardial α-adrenoreceptors. J Molec Cell Cardiol 14:185–188
49. Bikle DD, Murphy EW, Rasmussen H (1975) The ionic control of $1,25(OH)_2D_3$ synthesis in isolated chick renal mitochondria. The role of calcium as influenced by inorganic phosphate and hydrogen ion. J Clin Invest 55:299–304
50. Horiuchi N, Suda T, Sasaki S, Ogata E, Ezawa I, Sano Y, Shimazawa E (1975) The regulatory role of calcium in 25-OH-vitamin D_3 metabolism in chick kidney in vitro. Arch Biochem Biophys 171:540–548
51. Gasser AB, Morgan DB, Fleisch HA, Richelle LJ (1972) The influence of two diphosphonates on calcium metabolism in the rat. Clin Sci 43:31–45
52. Baxter LA, DeLuca HF, Bonjour JP, Fleisch HA (1974) Inhibition of vitamin D metabolism by ethane-1-hydroxy-1,1-diphosphonate. Arch Biochem Biophys 164:655–662
53. Goulding A, Cameron V (1978) Effects of diphosphonate on kidney calcium content and duodenal absorption of ^{45}Ca. Horm Metab Res 10:573–574
54. Harris DCH, Hammond WS, Burke TJ, Schrier RW (1986) Protective effect of verapamil on progression of experimental chronic renal failure. Kidney Intl 29:319A

Calcium Antagonists in Acute Renal Failure and in Hypertension

Calcium Antagonists in Ischemic Acute Renal Failure (Improvement of Function and Morphologic Damage by Nisoldipine)

B. Garthoff[1] and L. Hertle[2]

Summary

The effects of the calcium channel blocker nisoldipine on renal function after 60 min normothermic ischemia and contralateral nephrectomy were studied in male Wistar rats. Nisoldipine was given in a standard diet as well as one hour prior to ischemia. Survival, serum urea, serum creatinine, urine volume and creatinine clearance were used to test the effectiveness of the drug. Nisoldipine treatment resulted in the survival of all animals, improved renal function and reduced morphological alterations. Kidneys of nisoldipine-treated animals had significantly lower calcium tissue content. The beneficial effects of the drug in post-ischemic acute renal failure (ARF) may be attributed in part to direct effects on ischemic renal epithelial cells, presumably by inhibiting transmembrane calcium fluxes.

Introduction

It has been emphasized that intracellular calcium in vascular smooth muscle and renal epithelial cells could be involved in renal vasoconstriction and epithelial cell necrosis in renal ischemia. Likewise, it was assumed that calcium antagonists blocking transmembrane calcium influx would protect against the functional and cellular consequences of renal ischemia [11]. Burke and Schrier [1] have shown that increased cytosolic calcium is critically important in the maintenance of renal vasoconstriction and the development of cellular necrosis with subsequent tubular obstruction in ischemic acute renal failure induced by intrarenal infusion of norepinephrine. The present investigations address the situation in ischemic acute renal failure induced by 60 min clamping of the remaining kidney of uninephrectomized rats; the sequelae of intracellular calcium accumulation were looked at by recording functional parameters as well as histopathology (light and electron microscopy) and calcium determination in renal tissue during the first phases of acute renal failure.

Although the calcium channel blocking drugs share the ability to completely inhibit transmembrane calcium fluxes in smooth and cardiac muscle [5], it has

---

[1] Institute of Pharmacology, Bayer AG, P. O. Box 101709, D-5600 Wuppertal 1, FRG.
[2] Department of Urology, Ruhr University Bochum, D-4630 Bochum, FRG.

Nephrocalcinosis, Calcium Antagonists,
and Kidney
Ed. by K.-H. Bichler and W.L. Strohmaier
© Springer-Verlag Berlin Heidelberg 1988

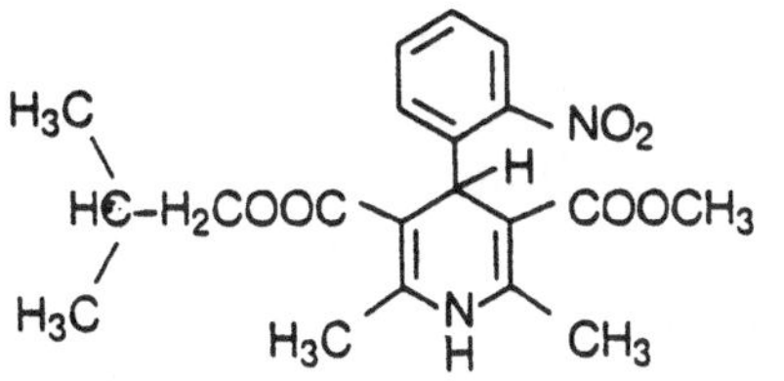

Fig. 1. Chemical structure of nisoldipine (Isobutyl methyl 1,4-dihydro-2,6-diethyl-4-(2-nitrophenyl)-3,5-pyridinedicarboxylate)

been noted that they are structurally dissimilar, possessing diverse pharmacological properties. Nisoldipine, a new dihydropyridine derivative (structure formula in Fig. 1), exerts potent relaxant effects on vascular smooth muscle. Experimentally, nisoldipine has demonstrated favorable effects on ischemic myocardium in the intact animal [8]. The drug also has an antihypertensive effect in various models of experimental hypertension [8]. The present study was undertaken to investigate whether nisoldipine could protect the rat kidney against ischemic damage.

Methods

All experiments were performed on male Wistar rats weighing 180 to 220 g, as we described earlier [7]. The experimental animals, divided into four groups (Table 1), were housed in individual metabolic cages and maintained on tap water and a standard pellet diet (ssniff-Versuchstier-Alleindiät), to which they had free access pre- and postoperatively. Four days prior to the study, the right kidney was removed through a small flank incision under ether anesthesia. For preparation of the left kidney, the animals were anesthetized with an intraperitoneal injection of sodium pentobarbital (Nembutal, Abbott) 40 mg/kg body weight (b. w.) and placed on a heated table that maintained body temperature between 37° C and 38° C. The kidney was exposed by a flank incision, freed from perirenal fat and left attached only by its pedicle. Temporary renal ischemia was induced by occlusion of the vascular pedicle with a microvascular clamp (Yasargil aneurysm clip, Aesculap, closing force 0.39 to 0.49 N) for 60 min. In sham operated animals, the

Table 1. Characteristics of experimental groups

Group	n	Protocol	Recovery period (days)
I	10	Sham-operated, untreated	14
II	9	Sham-operated, nisoldipine treated (prior and subsequent to sham operation)	14
III	33 (10/23)	Normothermic ischemia, untreated	14/3
IV	28 (10/18)	Normothermic ischemia, nisoldipine treated (prior and subsequent to ischemia)	14/3

Groups III and IV were subdivided; with 10 animals of each group followed 14 days and 23 animals of group III and 18 animals of group IV followed 3 days after ischemia.

kidneys were prepared identically except for clamping the vascular pedicle. The flank was left exposed for a full 60 min. After completion of surgery and recovery from anesthesia the rats were returned to their cages. Nisoldipine (Bayer AG) was given in the standard diet four days prior to, and during the three or fourteen days observation period after ischemia. The concentration of the drug in the diet was 300 ppm (mg/kg diet). Additionally, nisoldipine (10 mg/kg b.w.) was administered one hour before induction of ischemia by gavage.

Animals were examined daily and weighed at regular intervals. Serum creatinine and urea were determined just prior to ischemia (day 0) and on days 1 and 3 in all groups. On the same days as indicated above, 24 h urine samples were collected. Urine volume and creatinine levels were determined. Endogenous creatinine clearance was estimated from total 24 h creatinine excretion and from the serum creatinine value.

Improved survival was tested using Fisher's exact test [3, 9]. The remaining data from the different groups 1 or 3 days after ischemia were subjected to analysis of variance. This procedure was followed by Dunnett's test or Scheffe's test to identify the sources of differences [12]; in cases where variance-instability was detected data were subjected to logarithmic transformation. A p-value of less than 0.05 was considered to be significant. Results are expressed as means $\pm$ standard error of means (SEM), if not differently indicated.

For histopathology by light microscopy, the remaining kidney was removed, fixed in formaldehyde and sections were stained with HE according to standard procedures.

For electron microscopy, a special series was fixed by perfusion with cacodylate buffer (1.25% glutaric aldehyde, 1.25% p-glutaric aldehyde). In situ precipitation of calcium was done by use of a mixture containing glutaric aldehyde, veronal-acetate buffer $0.014\,M$, pH 7.1 with additional ammonium oxalate $(0.0125\,M)$.

Calcium content of renal tissue was estimated in another special series on six rats each, in which the remaining kidney was perfused in situ with ice-cold lanthanum buffer via aorta at the time intervals defined (4 days after uninephrectomy, 6 h, one day, and three days after ischemia). Half of the kidney was shock-frozen in liquid nitrogen and homogenized by a Braun dismembranator. Calcium content was determined by atomic absorption (Perkin Elmer 3020) in a preweighed specimen decomposed by hydrochloric acid and taken up in lanthanum buffer.

Results

Control values before ischemia (day 0) for all tested parameters (mean values $\pm$ standard deviations; no. = 110) were: serum urea 9.2 ± 1.6 mmol/l, serum creatinine 62 ± 15 µmol/l, urine volume 3.7 ± 1.6 µl/min and endogenous creatinine clearance 518.3 ± 175.7 µl/min.

Figure 2 and Tables 2–4 summarize the renal function changes following sham operation or ischemia in the 4 groups of animals. Sham operation without

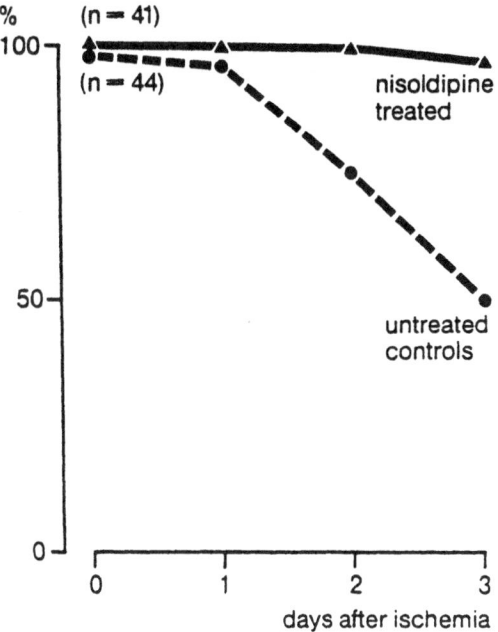

Fig. 2. Survival rate in acute ischemic renal failure. Fifty percent of the untreated animals died up to the third day after ischemia, whereas in the nisoldipine treated groups only one of forty-one animals died

and with nisoldipine treatment (groups I and II) had no detrimental effects on renal excretory function over a fourteen day observation period.

Rats subjected to 60 min normothermic ischemia without any protection (group III) developed an ARF with about one third of the animals being totally anuric on the first day after declamping. Most of these animals died two to six days after the ischemic insult, and in all cases deaths were due to uremia. All animals of this group had raised serum urea and creatinine concentrations and a diminished creatinine clearance on the first, third and seventh day after ischemia and the values had not returned to normal 14 days after ischemia. The peak rise in urea and creatinine occurred on day 3 after ischemia. By day 3 the surviving animals excreted large volumes of dilute urine, more than 200 per cent of the preischemic values (day 7: 27.2 ± 1.9 µl/min and day 14: 20.5 ± 2.9 µl/min; no. = 6).

All animals treated with nisoldipine prior and subsequent to normothermic ischemia (group IV) survived the insult (significantly different versus group III on day 3, $p < 0.0005$). Rats of group IV had significantly lower serum urea, serum creatinine, and higher creatinine clearance on the first and third day after ischemia than the untreated rats (group III). Animals treated with nisoldipine had a significantly higher urine volume on day 1 after ischemia than the untreated group. On days 7 and 14 after ischemia the nisoldipine treated animals had normal serum urea, serum creatinine and creatinine clearance values, whereas these values were still abnormal in the untreated rats (Fig. 3).

Histopathological investigation by light and electron microscopy revealed loss of brush border in the proximal tubular cells, alterations of the cell shape such as cuboidal or squamous type or revealed apical vesiculation or vacuolisation ("blebs") in addition to the loss of brush border in the untreated animals. Vacuolisation was especially prominent in the S_3-segment. Distal tubules, especially

Table 2. Serum creatine and creatinine clearance after sham operation or 60 min normothermic ischemia followed over 14 days. Number of surviving animals in brackets. Mean ± SEM

Experimental group	Serum creatinine (µmol/l)				Creatinine clearance (µl/min)			
Day	1	3	7	14	1	3	7	14
I ($n=10$)	57 ± 2 (10)	63 + 7 (10)	57 ± 2 (10)	61 ± 2 (10)	610 ± 26	609 ± 70	541 ± 46	581 ± 40
II ($n=9$)	56 ± 2 (9)	54 ± 1 (9)	61 ± 6 (9)	66 ± 5 (9)	547 ± 29	465 ± 66	425 ± 31	477 ± 43
III ($n=10$)	274 + 23 (9)	302 ± 33 (7)	171 ± 18 (6)	104 ± 13 (6)	49 ± 12	111 ± 19	199 ± 43	425 ± 92
IV ($n=10$)	170 ± 15 (10)	126 ± 31 (10)	68 ± 2 (10)	56 ± 1 (10)	144 ± 17	338 ± 44	454 ± 38	563 ± 52

For statistical evaluation refer to Tables 3 and 4.

Table 3. Serum creatinine (SCr), urine volume (U Vol) and creatinine clearance (CrCl) on the first day after sham operation or ischemia. Means ± SEM

Experimental group	SCr (µmol/l)	U Vol (µl/min)	CrCl
I	57 ± 2[a,**]	4.4 ± 0.4	610.4 ± 25.6[a,**]
II	56 ± 2[a,**]	5.2 ± 0.6	547.3 ± 28.7[a,**]
III	259 ± 11	5.3 ± 0.8	49.7 ± 10.8
IV	196 ± 14[a,**]	11.1 ± 1.1[a,**]	132.1 ± 17.1[a,**]

[a] Significantly different from group III (corresponding subgroups).
** $p < 0.01$.

Table 4. Serum creatinine (SCr), urine volume (U Vol) and creatinine clearance (CrCl) on the third day after sham operation or ischemia. Means ± SEM

Experimental group	SCr (µmol/l)	U Vol (µl/min)	CrCl
I	63 ± 7[a,**]	3.7 ± 0.3[a,**]	608.6 ± 69.6[a,**]
II	54 ± 1[a,**]	2.9 ± 0.7[a,**]	464.6 ± 66.2[a,**]
III	280 ± 23	13.0 ± 1.5	122.2 ± 20.0
IV	201 ± 21[a,**]	13.3 ± 0.2	224.7 ± 28.9[a,**]

[a] Significantly different from group III (corresponding subgroups).
** $p < 0.01$.

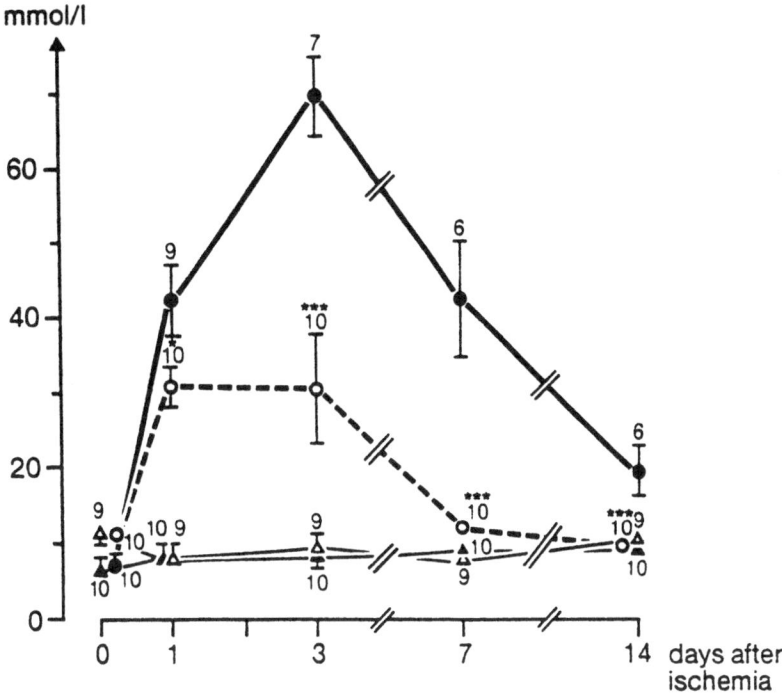

Fig. 3. Serum urea concentrations in uninephrectomized rats after sham operation or 60 min of normothermic ischemia: ▲ sham-operated, untreated (group I), △ sham-operated, nisoldipine treated (group II), ● ischemia, untreated (subgroup III), ○ ischemia, nisoldipine treated (subgroup IV). Means ± SEM; number of surviving animals given at symbols. Day 0: values before sham operation or ischemia. Raised serum urea levels throughout observation period in ischemia-untreated group with considerable mortality. No mortality and lower urea levels at all days after ischemia in nisoldipine-treated group. *p < 0.05; ***p < 0.001

partes rectae were extremely damaged; lumina were filled with desquamized epithela forming cylinder with calcium incrustation (Fig. 4 a, b).

In treated animals, "blebbing" was also observed, but apical vacuolisation was less severe and there was no desquamazation. Calcium precipitation could be seen in the proximal tubular cells of treated rats, however to a lesser extent. Calcium incrustation in distal tubules was limited to intracellular cristals, which were in the process of being diminished by increased lysosomal activity. In this sense, tubules of treated animals experienced a lesser calcium accumulation and an intense reparation activity.

The determination of calcium content in kidneys of untreated rats revealed that calcium content progressively increased after induction of acute renal failure. Calcium was almost doubled in renal tissue 3 days after ischemia. Although calcium content in renal tissue of treated rats increased slightly, the difference between values before and after clamping of the renal artery were not statistically significant. Kidneys of nisoldipine-treated animals had significantly lower calcium content than those of untreated, at least three days after ischemia (302 ± 58 versus 649 ± 69 µg/g wet weight).

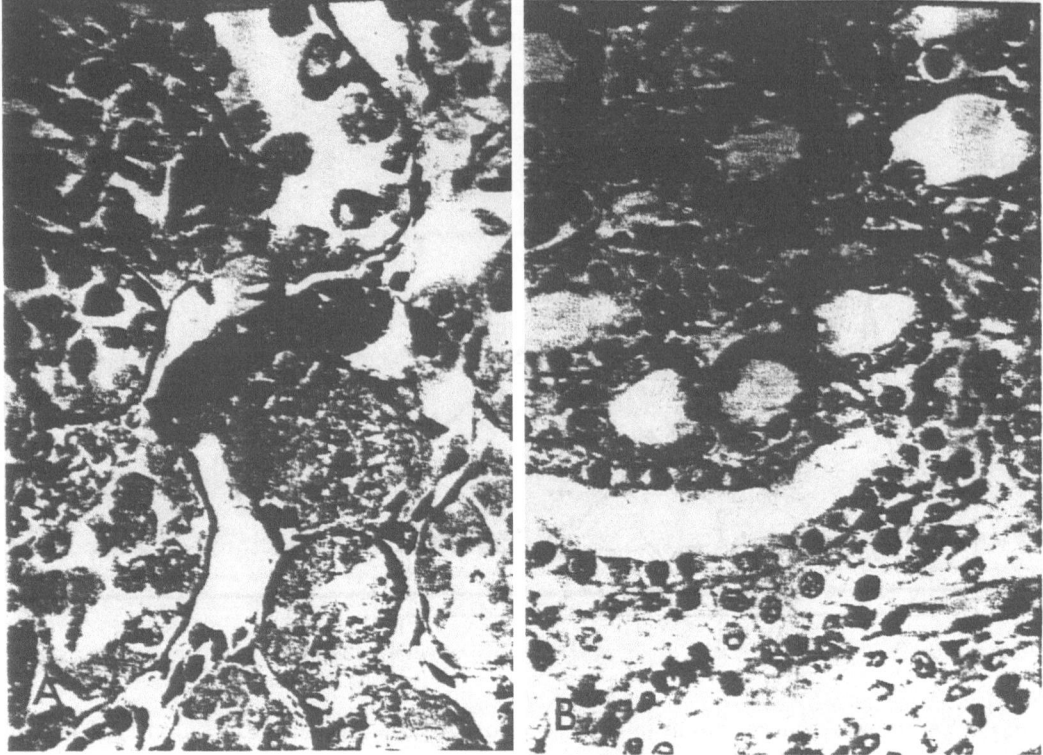

Fig. 4 A, B. HE-stained light micrographs from animals of the untreated group (A) and of the nisoldipine-treated group (B) at the third day after ischemia. Severe proximal tubular necrosis in the kidney subjected to ischemia without any protection. In contrast, widely patent tubular cells in the section from a well-protected nisoldipine-treated animal

Discussion

A period of 60 min normothermic ischemia after contralateral nephrectomy resulted in severe and sometimes lethal renal injury. Nisoldipine treatment before inducing ischemia, as well as continuing treatment in the period thereafter, was found to be highly effective in preventing uremic death and improving immediate and long term (14 days) renal function. The beneficial effects of calcium channel blockers on ischemically damaged tissues have been attributed to the blockade of deleterious calcium influx [6]. Although the cellular and molecular mechanisms leading to ischemic cell damage have not been clarified, substantial evidence exists that the influx of calcium plays a major role in mediating cellular injury [4]. Schanne et al. [10] showed that a range of membrane-active toxins only caused toxic cell death in the presence of normal extracellular calcium activity. When the extracellular calcium activity was reduced to that of normal intracellular activity, the toxins had little effect.

The biochemical abnormalities which may initiate cellular death during renal ischemia, thereby providing substrate for tubular obstruction, have not been defined. Burke and Schrier [1] showed that in renal ischemia a progressive increase in mitochondrial calcium accumulation occurs during reflow.

The present results confirm the previous findings by Burke et al. [2]. Although in their studies in an norepinephrine-induced dog model of acute renal failure renal tissue calcium was not different between untreated and verapamil-treated animals, substantial protection of mitochondrial function was observed as assessed by cellular respiration and calcium kinetics in vitro [2]. Interestingly, significant effects were not seen before 24 h after the ischemic insult, correlating to our findings on tissue content of rats. As demonstrated by our histopathological investigations, the most probable explanation is the occurrance of tubular necrosis. This cell death provides the debris for the development of tubular obstruction. It is conceivable that early cellular dysfunction is the consequence of the ischemic insult rather than the increased calcium concentration in mitochondria. However, administration of calcium antagonists such as nisoldipine protect from progression of ischemic injury to necrosis.

Although the results obtained in this study should be extrapolated with caution to the clinical situation, there is increasing appreciation of the potential of pharmacological calcium channel blockade in renal ischemia. Recently, it was reported [3] that the incidence of ARF after cardiovascular surgery dropped after calcium channel blockers were used pre- and intraoperatively to improve cardiac function. It would seem that these drugs had an unexpected but decidedly beneficial side effect.

References

1. Burke TJ, Schrier RW (1983) Ischemic acute renal failure – pathogenetic steps leading to acute tubular necrosis. Circ Shock 11:255–259
2. Burke TJ, Arnold PE, Gordon JA, Bulger RE, Dobyan DC, Schrier RW (1984) Protective effect of intrarenal calcium membrane blockers before or after renal ischemia. J Clin Invest 74:1830–1841
3. Editorial (1983) Promising agents limiting renal damage. JAMA 249:1987
4. Farber JL (1981) Minireview. The role of calcium in cell death. Life Sci 29:1289
5. Fleckenstein A (1977) Specific pharmacology of calcium in myocardium, cardiac pacemakers and vascular smooth muscle. Ann Rev Pharmacol Toxicol 17:149
6. Fleckenstein A (1983) Prevention by calcium antagonists of deleterious calcium overload: a principle of cardioprotection. In: Fleckenstein A (ed) Calcium antagonism in heart and smooth muscle, chap 3. Wiley, New York, pp 109–164
7. Hertle L, Garthoff B (1985) Calcium channel blocker nisoldipine limits ischemic damage in rat kidney. J Urol 134:1251
8. Kazda S, Garthoff B, Rämsch K-D, Schlüter G (1983) Nisoldipine. In: Scriabine A (ed) New drugs annual: cardiovascular drugs. Raven Press, New York, pp 243–258
9. Sachs L (1983) Angewandte Statistik, 6. Aufl. Springer, Berlin Heidelberg New York, S 288–290
10. Schanne FAX, Kane AB, Young EE, Farber JL (1979) Calcium dependence of toxic cell death: a final common pathway. Science 206:700
11. Schrier RW, Burke TJ, Conger JD, Arnold PE (1981) Newer aspects of acute renal failure. Proc 8th Int Congr Nephrol, Athens, pp 63–69
12. Wallenstein S, Zucker CL, Fleiss JS (1980) Some statistical methods useful in circulation research. Circ Res 47:1

Effect of Calcium Antagonist Diltiazem on Acute Renal Failure: Experiments on Animals and Clinical Studies

K. WAGNER[1] and H. H. NEUMAYER[1]

Influx of calcium ions into the cell represents a ubiquitous second messenger system [1] which is involved in a number of physiologic and pathophysiologic processes [2]. Apart from its role in the pathogenesis of arterial hypertension the contribution made by abnormal calcium homeostasis to cellular damage is more and more clearly appreciated [3]. It should be remembered that the resting extracellular calcium concentration is greater than the intracellular by a factor of 10^4. One of the principal consequences of damage to the cell membrane is a massive influx of calcium into the cell. The deleterious role of intracellular calcium accumulation following an ischemic insult has been well documented in the case of cardiac muscle and liver [3–5]. In the hope that they would also inhibit calcium influx into the damaged cell, calcium antagonists, which block socalled slow channels, have been studied in animal experiments on myocardial infarction and ischemic or toxic hepatocellular insults [5, 6]. Schrier [7] was the first to draw attention to the pathophysiologic significance of intracellular and intramitochondrial calcium accumulation for the development of acute renal failure (ARF). Animal experiments undertaken by the same research group demonstrated a protective effect of the calcium antagonist verapamil in norepinephrine induced renal failure. These experiments do not, however, exclude that the protection arose entirely from a direct effect on the vasculature with consequent improved perfusion, since simultaneous administration of norepinephrine and verapamil results in a minimal residual blood flow through the kidney [9]. In addition, all experiments were undertaken in anaesthetized animals, so that the effect of anesthesia on renal perfusion also has to be considered [10].

We have developed an animal model based on the chronically instrumented dog in which ARF can be induced in the unanesthetized state, thus dispensing with anesthetic artefacts. We have used this model to study the effect of a continuous calcium antagonist infusion on postischemic ARF.

Calcium antagonists of the dihydropyridine group (nifedipine etc) are not suitable for continuous infusion because of their photosensitivity. Continuous administration of verapamil carries not only the risk of cardiac side effects but also of drug accumulation if renal function is impaired [11]. On the other hand the calcium antagonist diltiazem is known not to accumulate in the presence of reduced renal function [12]. Furthermore the drug is stable in daylight and at room temperature. Also, diltiazem has few clinical side effects [13].

---

[1] Department of General Medicine and Nephrology, Klinikum Steglitz, Hindenburgdamm 30, D-1000 Berlin 45

Nephrocalcinosis, Calcium Antagonists, and Kidney
Ed. by K.-H. Bichler and W.L. Strohmaier
© Springer-Verlag Berlin Heidelberg 1988

We discuss below the results of animal experiments in which diltiazem was administered to chronically instrumented dogs with postischemic renal failure. The therapeutic avenues suggested by this animal work are then considered in the light of two randomised prospective trials of the effect of diltiazem on ARF after cadaveric renal transplantation.

Part I

Effect of the Calcium Antagonist Diltiazem in Chronically Instrumented Conscious Dogs

Animal Model. Beagles were submitted to unilateral nephrectomy and, after a 6 to 8 weeks adaptation and hypertrophy period, an electromagnetic flow probe and an occlusion cuff (inflatable plastic cuff) were applied to the contralateral renal artery. An aortic catheter was inserted to measure arterial blood pressure and to permit continuous infusion of diltiazem via a portable minipump. A healing period of eight days was allowed before further studies. A computer system monitored haemodynamic parameters at 30 s intervals and glomerular filtration rate (GFR) was measured by single shot techniques (inulin, ^{51}Cr – EDTA) [14]. Acute renal failure was initiated by temporary inflation of the occlusion cuff. The resulting complete cessation of renal blood flow (RBF) was confirmed by the electromagnetic sensor, by scintigraphy and by the occurrance of anuria. Statistical evaluation was by the Dixon & Mood test and the Mann-Witney U-test [15] (for details of the animal model see [16]).

Effect of the Calcium Antagonist Diltiazem on Renal Function

Methods. Diltiazem was infused for 24 h at a dose of 2.5, 5 or 10 µg/kg/min. Measurements were made before commencing and before ending the infusion.

Results

As expected there was a dose related fall in mean arterial blood pressure from 119 ± 2 mmHg to 105 ± 2 mmHg (at 10 µg/kg/min; $p < 0.05$) without an attendant rise in heart rate. Just as with other calcium antagonists there was a rise in plasma renin activity (PRA) from 3.5 ± 1 to 7.5 ± 1 ng AT I/ml/h (at the highest dose; $p < 0.05$). The GFR fell, but not significantly so, from 54 ± 6 ml/min to values between 38 ± 4 and 41 ± 4 ml/min, whereas RBF rose at the lowest dose of diltiazem from 151 ± 17 to 170 ± 12 ml/min. The middle dose had no effect on RBF and the highest dose reduced it to 123 ± 16 ml/min. Renal vascular resistance (RVR) was reduced at a dosage of 2.5 or 5 µg/kg ($p < 0.05$) (Fig. 1).

Effect of the Calcium Antagonist Diltiazem on Postischemic Renal Failure

Methods. ARF was induced by inflating the occlusion cuff for a period of 180 min. A control group of animals (A, n = 5) received an infusion of 0.9% NaCl

K. Wagner and H. H. Neumayer

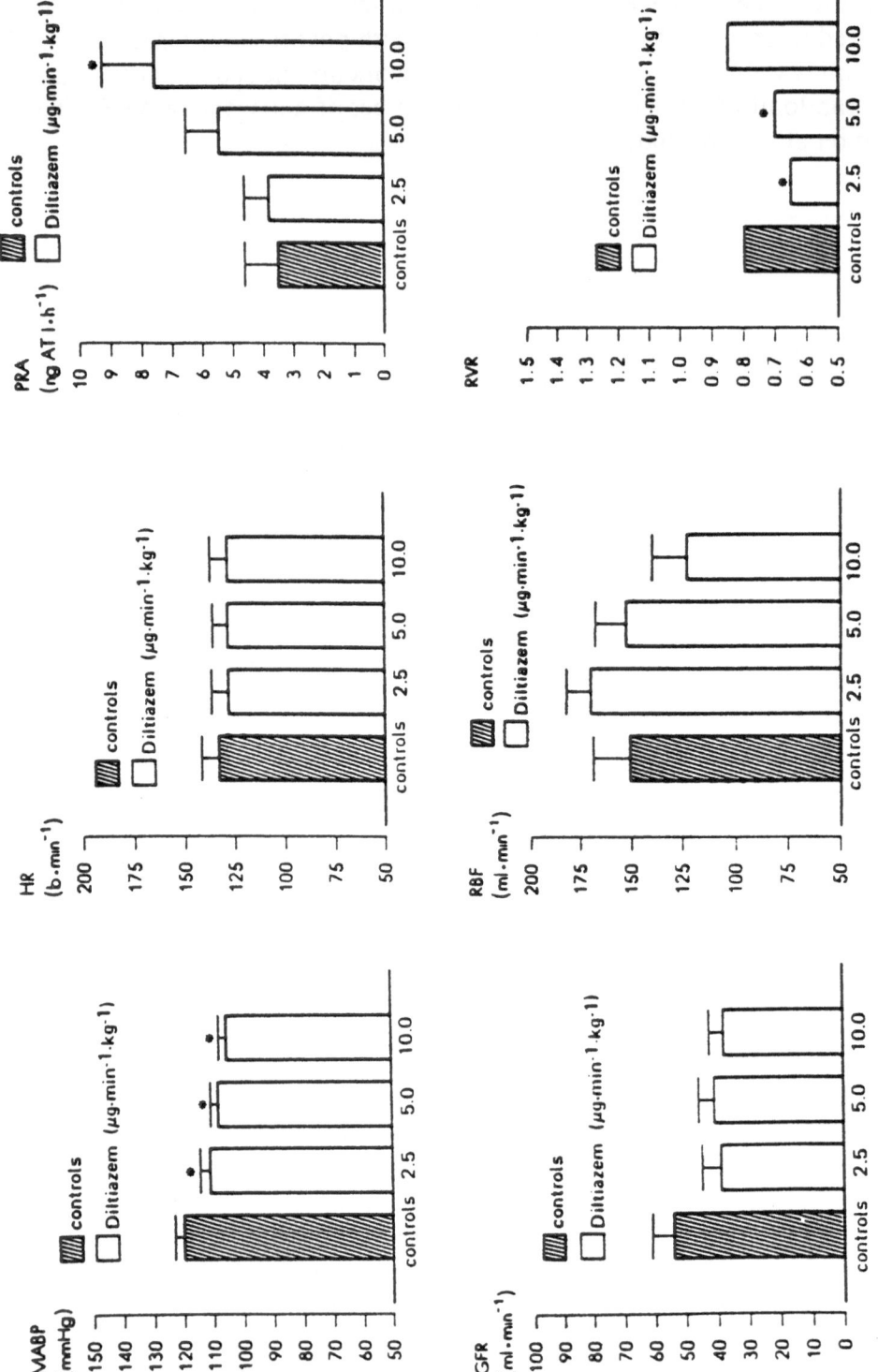

Fig. 1. Mean arterial blood pressure (*MABP*), heart rate (*HR*), plasma renin activity (*PRA*), glomerular filtration rate (*GFR*), renal blood flow (*RBF*), renal vascular resistance (*RVR*): effects of infusing Diltiazem at 2.5, 5, 10 μg/kg/min for 24 h (mean, SEM, * $p < 0.05$)

solution (5 ml/day) from the 3rd pre- to the 7th postischemic day. The post-treatment group (B, $n = 6$) received a continuous infusion of diltiazem at a dose of 5 µg/kg from immediately after ischemia until the 7th postischemic day. The full-treatment group (C, $n = 7$) received the same dose of diltiazem, but from the third day before ischemia until the seventh postischemic day. Renal function was determined on the day before ischemia, on the first, third and seventh postischemic days.

Results

In the control (A) and post treatment (B) groups GFR on the first postischemic day was reduced by $83 \pm 10\%$ and $76 \pm 3\%$ respectively. However, in the full treatment group (C) the fall in GFR was significantly less at only $52 \pm 11\%$ ($p < 0.05\%$). Whilst by the seventh postischemic day there was no difference between the GFR of the control group A and the post treatment group B, the GFR of the full-treatment group C was significantly higher ($p < 0.05$). In the control group A RBF fell by $27 \pm 5\%$ on the first postischemic day, whereas in both treated groups it rose by 29 ± 15 and 14 ± 13 respectively ($p < 0.05$) (Fig. 2). PRA rose significantly in the post treatment group until the seventh postischemic day but remained unchanged in the control group A. Parameters indicating an insult to tubular function (fractional sodium excretion, osmolar clearance, reabsorption of free water) were markedly less severely deranged in the full-treatment group C than in the control and post treatment groups A and B (Fig. 3).

Discussion of Animal Data

In our model a 180 min interruption of renal perfusion resulted in sublethal ARF. Even after ischemia of up to 360 min the dog kidney remains able to recover some function [17]. The sublethal course of our experimental ARF was further favoured by the protective effect of contralateral nephrectomy [18] and by the fact that the studies were undertaken in unanesthetized animals, thus obviating adverse effects of anesthesia on renal function [19].

Our studies demonstrate that administration of diltiazem was able to block any reduction in RBF on the first postischemic day in either treatment group. RBF was indeed raised above the preischemic level, in marked contrast to studies using verapamil which showed no effect on RBF in postischemic ARF in anesthetized dogs [20].

However, the improved perfusion due to diltiazem was limited to the initial phase of ARF. Over the following (3rd to 7th) days RBF fell below the starting value. At the same time PRA rose in both treatment groups, more markedly so in the post treatment group B. This may be explained as follows: It has been established that ischemia leads to a rise in the intracellular calcium concentration [21]. However, calcium influx into the juxtaglomerular cell is an inhibitory signal for renin secretion [22]. The reduction of this calcium influx by diltiazem would thus reduce the calcium-mediated inhibition of renin secretion, indirectly enhancing any secretion of renin in response to a suitable stimulatory signal. Following prostaglandin administration, by contrast, there is only a transient rise in PRA

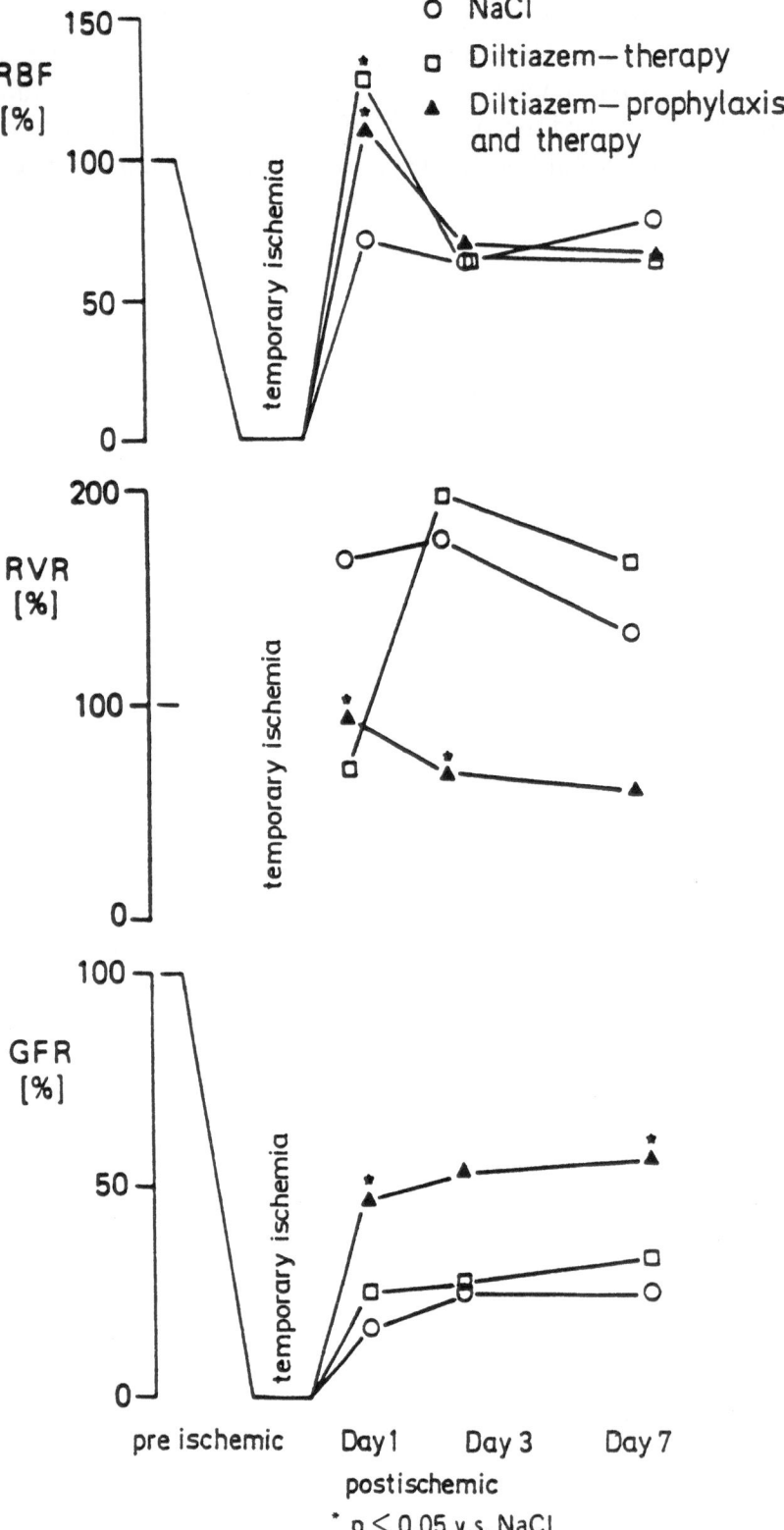

Fig. 2. Glomerular filtration rate (*GFR*), renal blood flow (*RBF*), renal vascular resistance (*RVR*). ○ = control group A (n = 5), □ = post treatment group B (n = 6) (Diltiazem treatment). Effect of infusing Diltiazem 5 μg/kg/min immediately after ischemia until the 7th postischemic day. △ = full treatment group C (n = 7, Diltiazem prophylaxis and treatment). Effect of infusing Diltiazem 5 μg/kg/min from 3rd preischemic to 7th postischemic day. (% change, mean, SEM, * p < 0.05)

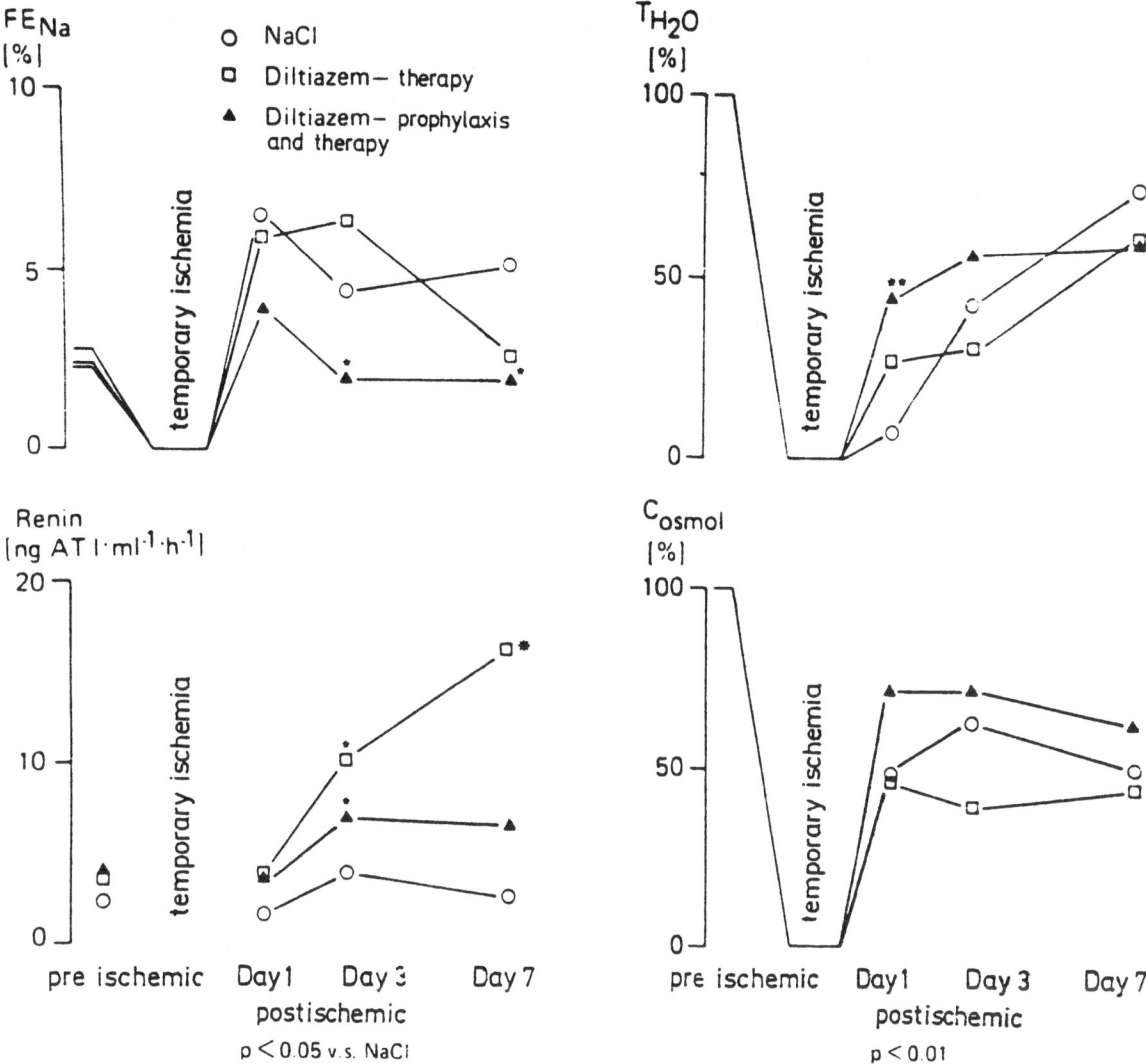

Fig. 3. Fractional sodium excretion (FE_{Na}), free water reabsorbtion (T_{H_2O}), plasma renin activity (PRA), osmolar clearance (C_{osm}). O = control group A (n = 5), □ = post treatment group B (n = 6) (Diltiazem treatment). Effect of infusing Diltiazem 5 μg/kg/min immediately after ischemia until the 7th postischemic day. Δ = full treatment group C (n = 7, Diltiazem prophylaxis and treatment). Effect of infusing Diltiazem 5 μg/kg/min from 3rd preischemic to 7th postischemic day. (% change, mean, SEM, * p<0.05)

immediately after ischemia [23]. As prostaglandins are without effect on calcium influx, an unreduced postischemic calcium influx remains able to inhibit renin secretion.

Our studies showed that administration of diltiazem before and after ischemia (full-treatment) has a protective effect on GFR. In the post treatment group B, which received diltiazem only after ischemia, this effect was lacking. A similar protective effect on GFR had previously been demonstrated in ARF induced by norepinephrine in anesthetized dogs [24]. However, that study has been criticized on the grounds, among others, that simultaneous administration of the vasodilator verapamil had reduced the vasoconstrictor effect of norepinephrine [9]. Even in the ARF induced by mechanical interruption of RBF the results remain controversial. In their pedicle-clamping rat model Malis et al. [9] found no effect of verapamil administration, whereas Goldfarb et al. [25] demonstrated a protective effect on GFR.

These differences may be explained by differences in calcium antagonist dose, by the duration of pre-treatment, which was occasionally 20 min [9] and by the brevity of post treatment, which amounted to only 20 min after ischemia [25]. No previous study has used prolonged administration of calcium antagonists, a more logical approach, since intracellular calcium levels rise to a maximum at 24 h after reperfusion [26].

In our model the deleterious effect of ischemia on tubular function was mirrored in a rise in fractional sodium excretion and in a fall of free water reabsorbtion. The less marked disturbances of tubular function seen in our full treatment group may be interpreted as tubular protection by diltiazem.

The Protective Effect of Diltiazem in Postischemic ARF May Therefore be Explained as Follows:

1. By inhibiting the postischemic influx of calcium into vascular smooth muscle cells, diltiazem improves renal perfusion. Calcium influx during the reperfusion phase results in a prolonged reduction in RBF [27] and contributes to reperfusion damage. However this vascular effect seems not to be the crucial one, since both treatment groups exhibited an initial increase in perfusion, yet only the full-treatment group C had an improved GFR. Furthermore, RBF fell in both treatment groups from the 3rd to the seventh postischemic day, yet the improved GFR of the full treatment group was preserved.

2. A direct protective effect of calcium antagonists on GFR has been demonstrated in micropuncture studies. The effect of angiotensin II on glomerular function was abolished by simultaneous administration of calcium antagonists [28]. Calcium antagonists enlarge the glomerular surface area by an effect on contractile mesangial elements [29].

3. It has been shown that subcellular events take place after ischemia that result in the liberation of free radicals which damage the cell membrane [30]. Some of these intracellular processes are calcium or Calmodulin dependent [31]. Thus a reduction in the rise of intracellular calcium during ischemic damage would reduce the evolution of toxic free radicals.

4. The most important mechanism of action of calcium antagonists on ARF must, however, be assumed to be a direct cellular one. In the reperfusion phase there is a cytosolic calcium influx in heart, liver and kidney cells [5, 6, 32, 33]. If the storage capacity of the endoplasmic reticulum is exceeded, calcium enters the mitochondria, interfering with oxidative phosphorylation [26, 32] and mitochondrial function in general [26]. Administration of calcium antagonists is able to interrupt this pathophysiologic chain by inhibiting or reducing the disastrous accumulation of calcium [34, 35]. This cell protective effect, which is independent of any vascular component, has also been confirmed in cell culture experiments. By adding calcium antagonists to the culture medium, the survival rate of tubular cells after anoxic damage could be augmented in a dose-dependent fashion [36].

Part II

Effect of the Calcium Antagonist Diltiazem on Acute Renal Failure After Renal Transplantation
Results of Two Randomized Prospective Trials

Introduction

The reported incidence of ARF following renal transplantation varies from 20 to 70% [37]. This variability can be explained by differences in the definition of ARF and in the immunosuppressive regime in use in individual units. The latter may have their own effect on renal function in the immediate wake of transplantation.

The ARF of renal transplantation is multifactorial in origin and its pathogenesis includes ischemic, toxic and immunologic causes. Apart from the general condition of the donor at the time of nephrectomy (hypotension, catecholamine therapy, administration of X-ray contrast media and nephrotoxic antibiotics) the effects of technique on ischemia time are worth mentioning. Ischemia time is in turn made up of: first warm ischemia time: the time required until the kidney can be perfused with Eurocollins' solution; cold ischemia time: the time for which the kidney is kept on ice pending transplantation; second warm ischemia time, during which vascular anastomoses are fashioned. Due to the modern technique of in situ perfusion and en bloc resection, first warm ischemia time can be reduced to under one minute. Among toxic factors active in the recipient cyclosporin A nephrotoxicity is of prime importance. Not only are there pressing psychological and organisational reasons for wanting to prevent ARF. There is good reason to suppose that the occurrence of ARF has a negative influence on long term graft survival [38, 39].

Prevention depends on strict adherence to short ischemia times, but this stipulation in turn reduces the organ exchange rate [40]. If the organ exchange falls there is a danger that lower standards of tissue type matching will be accepted and this in turn will affect the graft survival rate [41]. Because the circumstances leading to post-transplantation ARF are largely predictable, this form of human ARF is especially suitable for investigating treatment strategies. Contradictory results have been reported for the effect of mannitol [42, 43], or saralasin [44]. It was the aim of two randomized prospective trials to test the effect of the calcium antagonist diltiazem on post-transplantation ARF.

Methods

Trial I (Full Treatment). 42 consecutive transplants using locally harvested organs were included in the trial. Diltiazem was added to the perfusion fluid (Eurocollins' solution) in a dose of 20 mg/mgl in a random sequence. Prior to surgery the recipient received a bolus of diltiazem 0.28 mg/kg followed by a continuous infusion of 0.0022 mg/kg until the second postoperative day. Diltiazem was then given orally at 60 mg bd.

Trial II (After Treatment). 21 consecutive transplants using shipped organs from outside donor centers were included in the trial. In this trial only the recipient was treated with diltiazem in same way as in trial I.

Diltiazem was not given if the graft recipient witheld consent, had preexisting hypotension (resting recumbent systolic blood pressure < 90 mmHg) or had SA or AV block not treated by pacemaker.

During initial parenteral ditiazem treatment ECG and blood pressure were continuously monitored, the latter by Dynamap.

All patients received an immunosuppressive regime of parenteral cyclosporin A (5 mg/kg) preoperatively and this infusion was repeated on the first postoperative day. From the second to the seventh day cyclosporin A was given by mouth (12 mg/kg). Subsequent dosage was calculated so as to hold whole blood cyclosporin A levels between 400 and 800 ng/ml in the first two months and subsequently between 200 and 400 ng/ml. In addition all patients received low-dose steroids.

Post-transplant ARF was defined as no life supporting Kidney function, requiring at least one dialysis session on whatever indication during the first postoperative week.

GFR and RBF were determined by single shot methods (inulin, PAH, corrected for hematocrit) [14].

Whole-blood cyclosporin A levels were determined by RIA [45]. Renal biopsies were not performed routinely and were reserved for such clinical indications as deteriorating function, suspicion of rejection or cyclosporin A nephrotoxicity). Statistical evaluation was by χ^2 test (F-test) and by U-test for independent unpaired samples [15].

Results

The characteristics of donors and recipients are given in Table 1. There are no significant differences between the control and treatment groups in either trial: in particular there was no discrimination by ischemia factors.

There were no side effects such as hypotension or dysrhythmias in the diltiazem treated groups.

Trial I (Full Treatment) (Table 2). Nine patients (41%) in the control group (n = 22) developed ARF, compared to only 2 (10%, p < 0.05) in the diltiazem group (n = 20). Among the control group a total of 78 dialysis sessions were required as against 12 in the treated group (p < 0.05). Each dialysis patient in the control group required 3.5 ± 0.4 sessions, whereas the treated patients required only 0.6 ± 0.2 (p < 0.05). In addition the incidence of rejection episodes within the first month was reduced in the diltiazem group (p < 0.05).

Diltiazem-treated patients whose graft functioned initially had a significantly higher GFR than those of the control group (Day 7: 39 ± 1.4 versus 24 ± 0.7 ml/min; p < 0.05) and a significantly higher RBF (Day 7: 213 ± 8 versus 122 ± 7 ml/min; p < 0.05, Fig. 4). There was also a significant difference in plasma creatinine levels, 378 ± 21 μmol/l in the control group and 226 ± 9 μmol/l in the diltiazem group on day 7. The superior renal function among diltiazem-treated patients was also evident in a reduced biopsy rate (6 versus 14, p < 0.05).

Table 1. Characteristics of donors and recipients. Trial I (full treatment), Trial II (after-treatment). CVP = central venous pressure (mean, SEM)

Trial I Number of patients	(n)	Controls 22	Diltiazem 20
Donors			
Age	(yr)	37 ± 0.7	43 ± 0.7
Plasma creatinine	(mcmol/l)	104 ± 2	92 ± 2
Diuresis	(ml/h)	291 ±11	269 ±11
Recipients			
Age	(yr)	40 ± 0.5	41 ± 0.6
Duration of hemodialysis	(months)	38 ± 1	36 ± 1
Number of mismatches	(n)	2.9± 0.05	2.7± 0.08
Cold ischemia time	(h)	21 ± 0.2	18.6± 0.2
Warm ischemia time	(min)	37 ± 0.4	31 ± 0.4
Perfusion score of donor kidney (3 = excellent 2 = good 1 = bad)		2.7± 0.02	2.7± 0.02
CVP	(cmH$_2$O)	4.9± 0.1	5.5± 1

Trial II Number of patients	(n)	Controls 11	Diltiazem 10
Donors			
Age	(yr)	34 ± 1.3	34 ± 1.9
Plasma creatinine	(mcmol/l)	104 ± 3	102 ± 2
Diuresis	(ml/h)	384 ±27	324 ±24
Recipients			
Age	(yr)	48 ± 1.1	43 ± 1.6
Duration of hemodialysis	(months)	57 ± 6	31 ± 3
Number of mismatches	(n)	1.9± 0.4	11.7± 0.3
Cold ischemia time	(h)	25 ± 0.8	24.6± 1.1
Second warm ischemia time	(min)	39 ± 1.5	37 ± 2.3
Perfusion score of kidney (3 = excellent 2 = good 1 = bad)		2.8± 0.04	2.6± 0.06
CVP	(cmH$_2$O)	3.9± 0.3	5.1± 1

Diltiazem treatment resulted in significantly higher whole-blood cyclosporin A levels within the first 4 postoperative weeks ($p < 0.01$) and only a drastic reduction in cyclosporin A dosage by 30% resulted in subsequent blood levels similar to those seen in the control group (Fig. 5). The two groups also initially had widely different liver function values (GOT, GTP, bilirubin). However, this difference disappeared later, after reduction of cyclosporin A dose (Fig. 6).

Trial II (After Treatment) (Table 2). In the control group (n = 11) 5 patients (45%) developed ARF and in the treatment group (n = 10) there were 3 cases (30%). The control group required a total of 54, the treatment group 19 hemodialysis sessions. Patients in the control group developing ARF required an average of 10.8 ± 1.7 dialyses compared to 6.3 ± 1.9 in the treatment group before adequate renal function was installed. In the control group there were 0.72 ± 0.02 episodes of rejection per patient and in the diltiazem group 0.2 ± 0.01. Among

Table 2. Results of Trial I (full treatment), Trial II (after-treatment). ARF = acute renal failure, HD = hemodialysis. (mean, SEM, * = $p < 0.05$)

	Controls	Diltiazem
Number of patients (n)	22	20
Incidence of ARF	9/41%	2/10%*
Number of HD sessions (total)	78	12*
Number of HD sessions (per patient)	3.5 ± 0.4	0.6 ± 0.2*
Number of biopsies	14	6
Number of rejection episodes per patient (in first month)	0.5 ± 0.03	0.15 ± 0.02*

Results of Trial II (post-treatment)

	Controls	Diltiazem
Number of patients (n)	11	10
Incidence of ARF	5/45%	3/30%
Number of HD sessions (total)	54	19
Number of HD sessions (per patient)	4.9 ± 0.7	1.9 ± 0.4
Number of biopsies	15	9
Number of rejection episodes per patient (in first month)	0.7 ± 0.01	0.1 ± 0.01
Whole blood Cyclosporin A levels (ng/ml in first postop. week)	1078 ± 43	1735 ± 207

those patients whose graft functioned from the outset there was no detectable difference in the GFR between control and diltiazem groups. However, in the diltiazem group the GFR rose more rapidly and by the seventh postoperative day had reached a mean value of 49 ± 4.0 ml/min. At this time the control group had a mean GFR of 33 ± 3.3 ml/min. RBF was higher in the control than in the treatment group, being 357 ± 157 on the seventh day in the former but only 282 ± 12

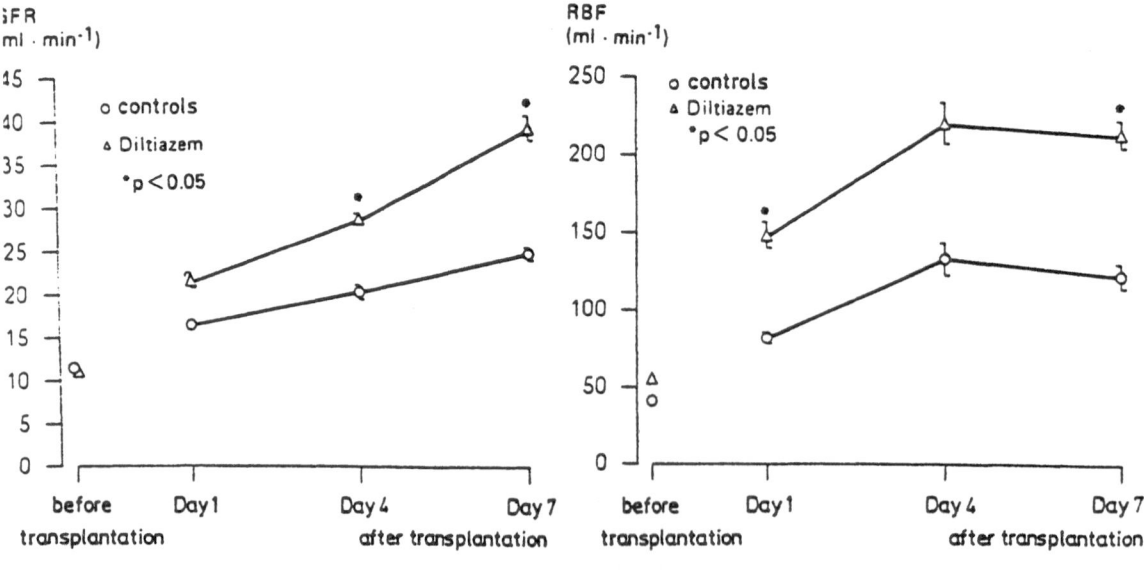

Trial I (full treatment)

Fig. 4. Glomerular filtration rate (*GFR*), renal blood flow (*RBF*). Trial I (full treatment), (mean, SEM, * p < 0.05)

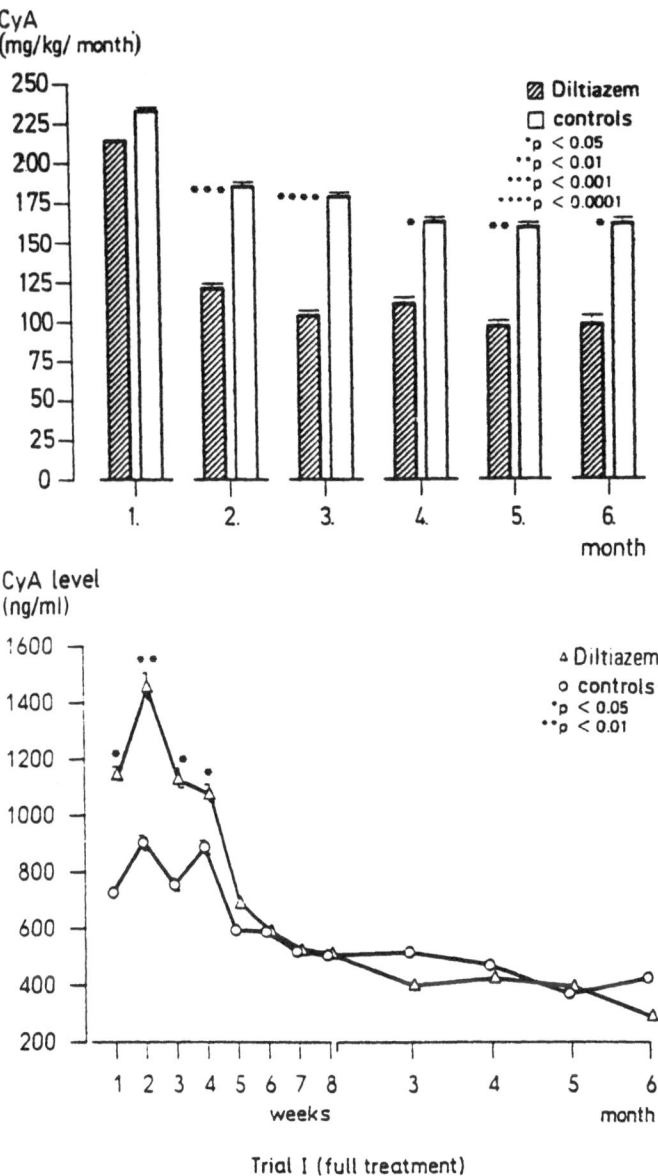

Fig. 5. Cyclosporin A dosage (*CyA*), whole blood cyclosporin A levels (*CyA level*). Trial I (full treatment), (mean, SEM, * p < 0.05, ** p < 0.01, *** p < 0.001)

ml/min in the latter (Fig. 7). The superior renal function of treated patients was apparent in their plasma creatinine levels, which were significantly lower than those of controls. On the seventh day the plasma creatinine of the controls (patients with ab initio function only) was 432 ± 57 µmol/l but only 124 ± 7 µmol in the Diltiazem group (p < 0.05). This significant difference in plasma creatinine levels was maintained between the two groups up until the third postoperative week (191 ± 16 versus 118 ± 8 µmol/l; p < 0.05). Whole blood Cyclosporin A levels were significantly higher between the first and third weeks in the Diltiazem group than in the controls (week 2: 1613 ± 123 versus 1082 ± 68 ng/ml). Within the first month the control group had undergone 15 biopsies and the patients treated with diltiazem 9.

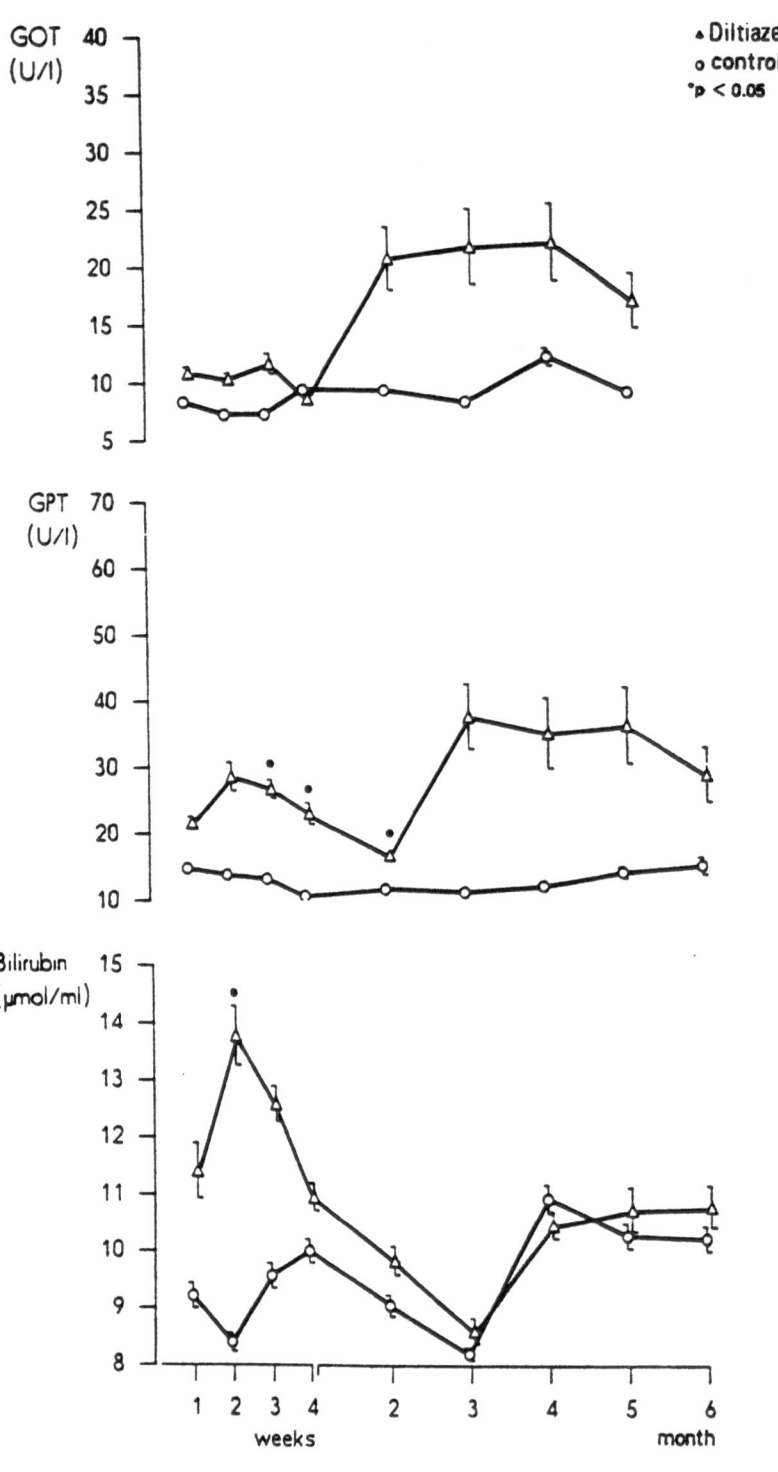

Trial I (full treatment)

Fig. 6. Glutamate-oxalate transaminase (*GOT*), glutamate-pyruvate transaminase (*GPT*), bilirubin. Trial I (full treatment), (mean, SEM, * p<0.05)

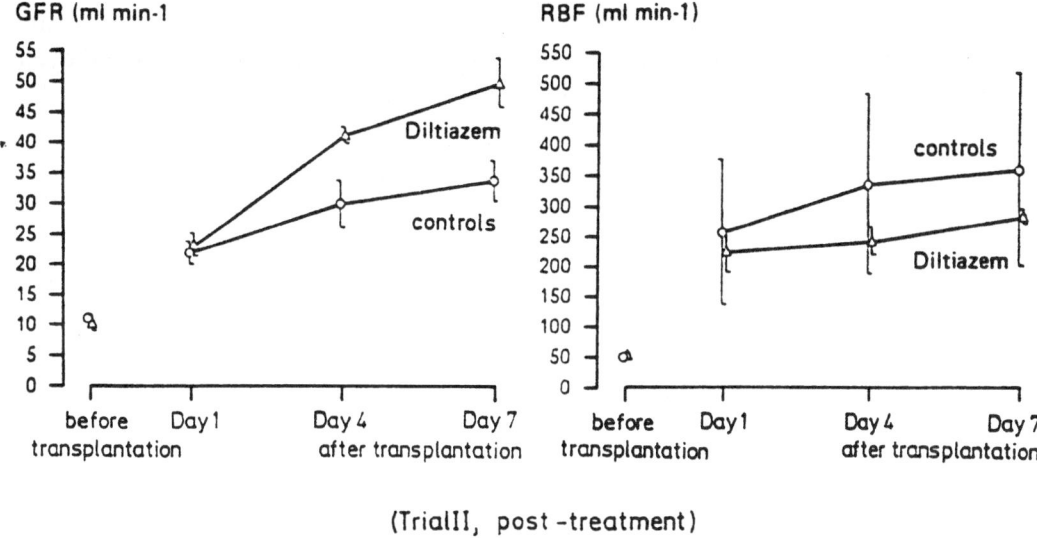

Fig. 7. Glomerular filtration rate (*GFR*), renal blood flow (*RBF*). Trial II (post treatment) (mean, SEM)

Discussion of Clinical Results

The hypothesis developed around the animal model, that the calcium antagonist diltiazem has a protective effect in postischemic ARF was fully confirmed by clinical studies on ARF after cadaveric renal transplants. Nevertheless it should be remembered that posttransplantation ARF is of multifactorial etiology. Correspondingly the effect of diltiazem manifested itself on the one hand in a reduced incidence of postoperative ARF and on the other in a reduced requirement for hemodialysis, an increased GFR and a lower plasma creatinine level. It was also demonstrated that preceding treatment of the graft recipient by adding diltiazem to the perfusion fluid (Eurocollins' solution) (full treatment) was more effective than treating the recipient alone (post treatment). The protective effect of the calcium antagonist diltiazem can be explained by its effect on the vascular and tubular components of the pathogenesis of ARF. As in the animal studies, vascular phenomena seem to be of secondary importance, since both in Trial I (full treatment) and in Trial II (post treatment) there was an improvement in GFR. Yet only in the full-treatment group was RBF enhanced. In the post treatment group RBF was less than in the controls. Since no parameters of tubular function were determined, a direct cytoprotective effect of diltiazem cannot be demonstrated from the clinical data: the reader is referred to the discussion of the animal data. It should be remembered that intracellular calcium accumulation peaks up to 24 h after reestablishment of the circulation [26]. Thus it should be clear why only giving a calcium antagonist prior to the ischemic insult is without clinical benefit. The same considerations may also explain why maximal benefit cannot be derived from giving the antagonist during the ischemic phase alone [47]. Further evidence for a primarily tubular site of action comes from the work of Atuk et al. [48]. These workers found that perfusing human kidneys with calcium antagonists reduced the functional and histological degree of tubular impairment.

Additional light is thrown on the improved function of diltiazem treated grafts by the finding that the simultaneous administration of diltiazem and cyclosporin A resulted in significantly higher whole-blood cyclosporin A levels. This discovery has recently been confirmed by other workers [49, 50]. Since both diltiazem [51] and cyclosporin A [52] are degraded in the liver, metabolism of diltiazem by Cytochrome-P-450-Oxidase [51] may reduce the residual capacity of that system to eliminate cyclosporin A. A 30% reduction in cyclosporin A dosage offers considerable economic advantages and achieves the same whole-blood levels.

The fact that renal function was significantly improved despite raised cyclosporin A levels poses the question whether the calcium antagonist diltiazem has a protective effect against the nephrotoxicity of cyclosporin A. The following findings need to be considered:

It is a matter of clinical experience that cyclosporin A nephrotoxicity is most marked in the presence of co-factors. Such co-factors include stimulation of the renin-angiotensin system [54], simultaneous use of antibiotics [55] and ischemic damage [56]. Protection by calcium antagonists against prior ischemic damage may thus reduce the sensitivity of the kidney to cyclosporin A nephrotoxicity.

In addition a direct protective mechanism may be at work. One histologic feature of cyclosporin A nephrotoxicity is microcalcification [57]. The latter may in turn be caused by calcium influx into the tubular cell and may be preventable by calcium antagonists. It has been shown in micropuncture studies that the tubular toxicity of cyclosporin A can be reduced by diltiazem administration [58]. A further possible explanation of reduced cyclosporin A nephrotoxicity in the presence of diltiazem may lie in reduced cyclosporin A uptake into the tubule cell if diltiazem is given at the same time. No such effect can be detected at the true site of action of cyclosporin A, the lymphocyte [59].

Raised whole blood levels of cyclosporin A are one possible explanation for the significantly fewer rejection episodes seen in diltiazem treated patients. Furthermore it has been suggested by van Es et al. [60] that ischemic tissue damage may lead to enhanced antigen expression and may trigger a rejection reaction. Protection against such ischemic damage would reduce the expression of antigen and with it the incidence of rejection episodes.

Summary

In animal experiments the calcium antagonist diltiazem has a protective effect against postischemic renal failure. This effect may result not only from improved organ perfusion after ischemia but, more importantly from a protective effect on the tubule. Such a cytoprotective effect can be explained by prevention of the known deleterious effects of intracellular calcium accumulation in the reperfusion phase.

A similar effect could be observed in clinical studies of posttransplantation ARF, although the latter is of mixed etiology. In addition diltiazem treatment of renal transplant patients offers both the economic advantage of reduced cyclosporin A dosage and protection against cyclosporin A nephrotoxicity.

References

1. Rasmussen H (1986) The calcium messenger system. N Engl J Med 314:1094–1101
2. Rasmussen H, Zawalich W, Kojima I (1985) Ca and cAMP in the regulation of cell function. In: Marme D (ed) Calcium and cell physiology. Springer, Berlin Heidelberg New York Tokyo, pp 1-17
3. Fleckenstein A (1971) Pathophysiologische Kausalfaktoren bei Myokardnekrose und Infarkt. Z f Inn Med 52:133–143
4. Farber JL, Mofty EL (1975) The biochemical pathology in liver cell necrosis. Am J Pathol 79:237–250
5. Fleckenstein A, Fleckenstein-Grün G, Frey M, Zorn J (1987) Future directions in the use of calcium antagonists. Am J Cardiol 59:177B–187B
6. Chien KR, Abrams J, Pfau RG, Farber JL (1977) Prevention by chlorpromazine of ischemic liver cell death. Am J Pathol 88:539–557
7. Schrier RW (1982) Acute renal failure. Jama 247:2518–2525
8. Burke ThJ, Arnold PE, Schrier RW (1982) Effect of calcium inhibition on norepinephrine induced acute renal failure. In: Eliahou HE (ed) Acute renal failure. Sibbley, London
9. Malis ChD, Cheung JY, Leaf A, Bonventre JV (1983) Effects of verapamil in models of ischemic acute renal failure in the rat. Am J Physiol 245:F735–742
10. Applegate CW, Gutman RA (1976) Renal intracortical blood flow distribution, function and sodium excretion in response to saline loading of anesthetized and unanesthetized dogs. Pflügers Arch 366:125–130
11. Henry PD (1980) Comparative pharmacology of calcium antagonists: nifedipine, verapamil and diltiazem. Ann J Cardiol 46:1047–1057
12. Pozet N, Brazier JL, HadjiAissa A, Kenfer D (1983) Pharmacokinetics of diltiazem in severe renal failure. Eur J Clin Pharmacol 24:635–638
13. Borchard U (1986) Neue therapeutische Ansätze zur Behandlung der arteriellen Hypertonie. In: Rosenthal J (Hrsg) Arterielle Hypertonie. Springer, Berlin Heidelberg New York Tokyo, S 766–796
14. Hall JE, Guyton AC, Farr BM (1977) A single-injection method for measuring glomerular filtration rate. Am J Physiol F 232:72–76
15. Sachs L (1975) Angewandte Statistik. Springer, Berlin Heidelberg
16. Wagner K, Neumayer HH, Schultze G, Schwietzer G, Schudrowitsch L, Ruf W, Molzahn M (1983) Influence of prostaglandin A_1 on renal filtration, hemodynamics and excretion. Renal Physiol 6:186–196
17. Balint P, Szöcs E (1976) Intrarenal hemodynamics following temporary occlusion of the renal artery in the dog. Kidney Int 10:128–134
18. Finn WF (1980) Enhanced recovery from postischemic acute renal failure. Micropuncture studies in the rat. Circ Res 46:440
19. Vatner SF (1974) Effect of hemorrhage on regional blood flow distribution in dogs and primates. J Clin Invest 54:225–230
20. Wait RB, White G, Davis JH (1983) Beneficial effect of verapamil on postischaemic renal failure. Surgery 94:276–282
21. Farber JL (1982) The role of calcium in cell death. Life Sci 29:1289–1295
22. Chun SP, Dae SH, Fray J (1981) Calcium in the control of renin secretion. Ca influx as an inhibitory signal. Am J Physiol 240:F70–F74
23. Neumayer HH, Wagner K, Groll J, Schudrowitsch L, Schultze G, Molzahn M (1985) Beneficial effect of long-term prostaglandin E_2 infusion on the course of postischemic acute renal failure. Renal Physiol 8:159–168
24. Burke Th, Arnold PE, Gordon JA, Bulger RE, Dobyan DC, Schrier RW (1984) Protective effect of intrarenal calcium membrane blockers before and after renal ischemia. J Clin Invest 74:1830–1840
25. Goldfarb D, Iaina A, Eliahou HE (1983) Beneficial effect of verapamil in ischemic acute renal failure in the rat. Proc Soc Exp Biol Med 172:389–392
26. Wilson DR, Arnold P, Burke Th, Schrier RW (1984) Mitochondrial calcium accumulation and respiration in ischemic acute renal failure in the rat. Kidney Int 25:519–526
27. Van Neuten JM, Manhoutte PM (1980) Improvement of tissue perfusion with inhibitors of calcium ion influx. Biochem Pharmacol 29:479–485

28. Ichikawa I, Miele JF, Brenner BM (1979) Reversal of renal cortical actions of angiotensin II by verapamil and manganese. Kidney Int 16:137–147
29. Martin-Dupont L, Cambar J, Brothier JP (1984) Effects of verapamil – a calcium inhibitor – on the vasomotor response of cortical and juxtamedullary glomeruli of human kidney. Renal Physiol 7:71–77
30. McCord JM, Roy RS (1982) The pathophysiology of superoxide: roles in inflammation and ischemia. Can J Physiol Pharmacol 60:1346–1352
31. DeMartino GN, Kuers K (1981) Two calcium-dependent, calmodulin-stimulated proteases from rat liver. Fed Proc Am Soc Exp Biol 40:1738
32. Schrier RW, Arnold PE, Burke ThJ (1982) Alterations in mitochondrial respiration and calcium movements in norepinephrine induced acute renal failure. In: Eliahou HE (ed) Acute renal failure. London
33. Burke ThJ, Schrier RW (1983) Ischemic acute renal failure: pathogenetic steps leading to acute tubular necrosis. Circ Shock 11:255–259
34. Wagner K, Schultze G, Molzahn M, Neumayer HH (1986) The influence of long-term infusion of the calcium antagonist Diltiazem on postischemic acute renal failure in conscious dogs. Clin Wochenschr 64:135–140
35. Neumayer HH, Wagner K (1986) Neue Aspekte zur Pathogenese des akuten Nierenversagens und mögliche therapeutische Konsequenzen. Niere und Hochdruckkrankheiten 15:235–250
36. Schwertschlag U, Schrier RW, Wilson P (1986) Beneficial effects of calcium channel blockers and calmodulin binding drugs on in vitro renal cell anoxia. J Pharm Exper Ther 238:119–124
37. Cao SI, Zalneraitis BP, Franklin C, Bradley JW (1985) The influence of acute tubular necrosis on kidney transplant survival. Transplant Proc 27:16–17
38. Keown PA, Stiller CR, Wallace AC, McKenzie FN, Wall W (1985) Cyclosporine nephrotoxicity: exploration of the risk factors and prognosis of the renal injury. Transplant Proc 17:247–253
39. Pichlmayr R, Wonigeit K, Ringe B, Neuhaus P, Frei V, Offner G, Brodehl J, Mihatsch J (1985) Sandimmun in renal transplantation. Basel
40. Eurotransplant Report 1985 (1986) Eurotransplant Foundation, Leyden
41. Opelz G, Najarian JS, Terasaki PI (1978) Prediction of long-term kidney transplant survival rates by monitoring early graft function and clinical grades. Transplantation 29:245–253
42. Weimar W, Geerling W, Bijnen AB, Obertrop H, van Urk H, Lameijer LDF, Wolff ED, Jeekel J (1983) A controlled study on the influence of mannitol on immediate renal function after cadaver donor kidney transplantation. Transplantation 35:99–104
43. Kaplan MP, Toledo-Pereyra LH, Pietroski R, Rosenberg JC, Allaban RD (1986) Effect of furosemide and/or mannitol on the immediate function of preserved cadaver kidneys. Transplant Proc 17/3:504–505
44. Huland H, Bause HW, Clausen C, Doehn N (1983) The influence of an angiotensin II antagonist, saralasin, given before donor nephrectomy, on kidney function after transplantation. Transplantation 36:139–143
45. Donatsch P, Abisch E, Homberger M, Traber R, Trapp M, Voges R (1981) A radioimmunoassay to measure cyclosporin-A in plasma and serum samples. J Immunoassay 2:19–32
46. Hull RW, Hasbargen JA (1985) No clinical evidence for protective effects of calcium-channel blockers against acute renal failure. N Engl J Med 313:1477–1478
47. Duggan VA, MacDonald GJ, Charlesworth JA, Pussel BA (1985) Verapamil prevents posttransplant oliguric renal failure. Clin Nephr 24:289–291
48. Atuk NO, Mihindu J, Sturgill BC, Teates CD, Westervelt FB, Rudolf L (1982) Protection of perfused human kidney for transplantation by verapamil. Clin Res 30:440A
49. Pochet JM, Pirson Y (1986) Cyclosporin-diltiazem interaction. Lancet I:979
50. Grino JM, Castelao AM, Alsina J (1986) Influence of diltiazem on cyclosporin clearance. Lancet I:1387
51. Sugihara J, Sugawara Y, Ando H, Harigaya S, Etoh A, Kohno K (1984) Studies on the metabolism of diltiazem in man. J Pharmacobio Dynamics 7:24–32
52. Freeman DJ, Laupacis A, Keown P, Stiller C, Carruthers G (1984) The effect of agents that alter drug metabolizing enzyme activity on the pharmacokinetics of cyclosporin-A. Ann R Coll Physns Surg Can 17:301–310

53. Neumayer HH, Wagner K (1986) Diltiazem and economic use of cyclosporin. Lancet II:523
54. Whiting PH, Cunningham CH, Thomson AW, Simpson BG (1984) Enhancement of high dose Cyclosporin-A toxicity by frusemide Biochem Pharm 33:1075–1078
55. Whiting PH, Simpson BG, Thomson AW (1983) Nephrotoxicity of cyclosporin in combination with aminoglycoside and cephalosporin antibiotics. Transplant Proc 15:2702–2703
56. Devineni R, McKenzie N, Duplan J, Keown P, Stiller C, Wallace AC (1983) Renal effects of cyclosporine: clinical and experimental observations. Transplant Proc 15:2695–2698
57. Mihatsch MJ, Thiel G, Spichtin HP, Oberholzer M, Brunner FP, Zollinger HU, Loertscher R (1983) Morphological findings in kidney transplants after treatment with cyclosporine. Transplant Proc 15:2821–2835
58. Gutsche H (1986) personal communication
59. Naginemi Ch, Yanagawa N, Misra B, Lee D (1987) Cyclosporin-A – calcium channel interaction: a possible mechanism for nephrotoxicity. Transplant Proc (in press)
60. van Es A, Hermans J, van Bockel JH (1983) Effect of warm ischemia time and HLA(A,B) matching on renal cadaveric graft survival and rejection episodes. Transplantation 36:255–258

Renal Function in Hypertensive Patients and Calcium Entry Blockers

G. Leonetti[1] and A. Zanchetti[1]

Calcium antagonists are employed in three different clinical conditions, i.e., coronary insufficiency, cardiac failure and arterial hypertension. In this presentation we report the results of our studies using calcium antagonists in the treatment of hypertensive patients and the concomitant effects on renal function.

In spite of the fact that blood pressure elevation in the majority of hypertensive patients is due to a rise in total peripheral resistances (and only in a minority to an increase in cardiac output), in the past the use of direct acting vasodilating drugs in monotherapy has been very limited. This was due to blood pressure reduction during treatment with direct acting vasodilating drugs which was accompanied by the signs and symptoms of reflex activation of the sympathetic nervous system. As cardiac output and peripheral vascular resistances increased concomitant with water and sodium retention as body fluid volume expanded, the antihypertensive action of direct acting vasodilating drugs was counterbalanced in part by the rise in the incidence of subjective side effects such as headache, tachycardia, flushing, etc. [1, 2].

Since the introduction of calcium antagonists, animal experiments have shown that the antihypertensive effect of calcium entry blockers is accompanied by a natriuretic and diuretic effect and by minor and transient symptoms of reflex activation of the sympathetic nervous system [3–5]. For these reasons we have investigated in hypertensive patients not only the antihypertensive action but also the renal effects and the degree of reflex activation of the sympathetic nervous system (these two effects, activation of the sympathetic nervous system and sodium and water balance, can be either independent of or dependent on each other). All the patients we have investigated were hospitalized in a metabolic ward to maintain a constant intake of sodium throughout the studies. This could only be obtained for short periods, therefore the studies we report are either acute or of short duration.

The first study we performed (Table 1) was the evaluation of the acute antihypertensive and renal effects of nifedipine (10 mg) and verapamil (160 mg) in single oral administration while the patients were kept on a constant daily sodium intake of 100 mEq [6]. The study was of 6 h duration during which the patients were kept supine. Both nifedipine and verapamil, with minor differences, caused significant blood pressure reductions when compared to placebo, while only nifedipine caused a mild and transient rise in heart rate. With regard to renal func-

[1] Istituto di Clinica Medica Generale e Terapia Medica, Università di Milano, Centro di Fisiologia Clinica e Ipertensione, Ospedale Maggiore, I-Milano, Italy.

Nephrocalcinosis, Calcium Antagonists, and Kidney
Ed. by K.-H. Bichler and W.L. Strohmaier
© Springer-Verlag Berlin Heidelberg 1988

Table 1. Acute effects of nifedipine and verapamil in hypertensive patients

	Placebo	Nifedipine	Verapamil
SBP (mmHg)	172 ± 6	$150 \pm 8*$	$153 \pm 6*$
DBP (mmHg)	118 ± 4	$100 \pm 5*$	$103 \pm 5*$
HR (b/min)	72 ± 3	78 ± 2	72 ± 3
Ur. Na$^+$ (mEq/6 h)	73 ± 34	$142 \pm 41*$	90 ± 28
Ur. Vol. (ml/6 h)	547 ± 246	$1032 \pm 399*$	670 ± 300
GFR (ml/min)	95 ± 7	102 ± 6	98 ± 9

Supine systolic (SBP) and diastolic blood pressure (SBP), heart rate (HR), urinary sodium excretion (Ur Na$^+$), urine volume (Ur.Vol.), and glomerular filtration rate (GFR) in 14 hypertensive patients in 6-h periods after oral administration of placebo, nifedipine (10 mg) and verapamil (160 mg) in different days (means $\pm$ SE).
* $p < 0.01$ active vs placebo.

Table 2. Effects of acute and repeated doses of felodipine in hypertensive patients

	Placebo	1st day	7th day
SBP (mmHg)	165 ± 5	$145 \pm 4**$	$138 \pm 5**$
DBP (mmHg)	106 ± 4	$93 \pm 3**$	$92 \pm 2**$
HR (b/min)	78 ± 2	$85 \pm 3*$	83 ± 2
GFR (ml/min)	88 ± 9	88 ± 7	92 ± 6
RPF (ml/min)	616 ± 36	$793 \pm 67*$	817 ± 79

Supine systolic (SBP) and diastolic blood pressure (DBP), heart rate (HR), glomerular filtration rate (GFR), and renal plasma flow (RPF) during placebo and the 1st and 7th day of felodipine administration (means $\pm$ SE).
* $p < 0,05$, ** $p < 0.01$ active vs placebo.

tion, nifedipine and verapamil in comparison to placebo caused an increase in water and sodium excretion which reached statistical significance only after nifedipine administration. Glomerular filtration rate, calculated as creatinine clearance, was not modified by either calcium antagonist, which suggests a direct action of calcium entry blockers on tubule cell sodium reabsorption as Di Bona [10] has shown in animal experiments.

In a second metabolic study (Table 2) we have investigated the effects of repeated oral doses of felodipine, a new calcium antagonist of the dihydropyridine group, on blood pressure and renal function [7]. The dose employed was 10 mg bid and the duration of the study was 7 days. The antihypertensive effect was alrady evident during the first day of felodipine administration and it was only slightly greater after 7 days of repeated oral administrations. Renal plasma flow, calculated as clearance of para-aminohippuric acid, was significantly increased during felodipine administration, while glomerular filtration rate was unchanged for the whole duration of the study. The cumulative urinary sodium output was positive, that is there was a loss of about 150 mEq in 7 days; also the cumulative urinary potassium output was positive, but of minor degree, that is about 50 mEq in 7 days. This study shows that felodipine lowers blood pressure while causing

Table 3. Dise-response action of felodipine on blood pressure and renal function

	Placebo	12.5	25.0	50.0 mg tid
SBP (mmHg)	181± 5	142± 4*	138± 5*	136± 4*
DBP (mmHg)	109± 2	91± 3*	89± 3*	85± 3*
GFR (ml/min)	92± 8	90± 9	84± 7	79± 8
Ur.Na$^+$ (mmol/24 h)	171± 16	152± 18	159± 17	139± 16
Ur.Vol. (ml/24 h)	1571±139	1477±201	1583±165	1602±122

Systolic (*SBP*) and diastolic (*DBP*) blood pressure, glomerular filtration rate (*GFR*), daily urinary sodium excretion (*Ur.Na$^+$*), and urine volume (*Ur.Vol.*) during administration of three different doses of felodipine (means±SE).
* $p < 0.01$ active vs placebo.

a mild natriuretic and diuretic effect during repeated oral administration over 7 days.

In another metabolic study (Table 3) we have again investigated the antihypertensive and renal effects of felodipine, but employing doses greater than that administered in the previous study: the doses were 12.5, 25, and 50 mg tid [8]. The greatest blood pressure reduction was already evident at the lowest dose of 12.5 mg tid and only minor blood pressure reductions were obtained by increasing felodipine doses to 25 and 50 mg tid. However, by employing higher doses we have not observed a natriuretic effect in this study, but on the contrary a trend toward sodium retention and also a mild reduction in creatinine clearance.

On the whole this study shows that by administering felodipine at higher doses there is only a slight further blood pressure reduction, but at the same time the hemodynamic changes at the renal level (as in glomerular filtration rate reduction) can counterbalance the natriuretic and diuretic action of the calcium antagonist at the tubule level. This indicates that felodipine has different dose-response curves for antihypertensive and natriuretic effects and with increasing doses the cost-benefit ratio worsens as the greater blood pressure reduction is not accompanied by the natriuretic and diuretic effects.

Verapamil [9] was administered for 10 days in another group of hypertensive patients kept on a constant sodium intake of 100 mEq/day (Table 4). Verapamil significantly decreased systolic and diastolic blood pressure 5 and 10 days after

Table 4. Antihypertensive and renal effects of repeated verapamil administration

	Placebo	5th day	10th day verapamil
MAP (mmHg)	133± 5	120± 5*	114± 3*
HR (b/min)	79± 3	78± 3	77± 2
Ur.Na$^+$ (mEq/24 h)	107±15	113±15	108±12
Ur.Vol. (ml/24 h)	910±65	1080±85	1032±72

Supine mean arterial pressure (*MAP*), heart rate (*HR*), urinary sodium excretion (*Ur.Na$^+$*) and urine volume (*Ur.Vol.*) during the last day of placebo and the 5th and 10th day of repeated verapamil administration (means±SE).
* $p < 0.01$ vs placebo.

repeated oral administration. Despite a blood pressure reduction the cumulative urine volume and sodium excretion after 5 and 10 days of treatment were unchanged when compared to pretreatment values. This confirms that during repeated verapamil administration, blood pressure reduction is not accompanied by sodium and water retention in contrast to the action of direct vasodilating drugs such as hydralazine, guancydine, and minoxidil in similar experimental conditions.

In conclusion from our studies we can say that:

1. Calcium antagonists are effective antihypertensive agents
2. The acute administration of calcium antagonists causes a natriuretic and diuretic effect (more evident with the dihydropyridine derivative).
3. Glomerular filtration rate is unchanged and renal plasma flow is increasing during calcium antagonist therapy.

References

1. Gross F (1977) Drugs acting on arteriolar smooth muscles (vasodilator drugs). In: Antihypertensive agents. Springer, Berlin Heidelberg New York, pp 397–476
2. Koch-Weser J (1979) Vasodilator drugs in treatment of hypertension. Arch Int Med 133:1017–1019
3. Yamaguchi I, Ikezawa K, Takada T, Kyomoto H (1974) Studies on a new 1,5-benzothiazepine derivative (CRD-40). VI. Effects on renal blood flow and renal function. Jp J Pharmacol 24:511–522
4. Brown B, Churchill P (1983) Renal effects of methoxyverapamil in anesthetized rats. J Pharmacol Exp Ther 225:373–377
5. Dietz JR, Davis JO, Freeman RH, Villareal D, Echtenkamp SF (1983) Effects of intrarenal infusion of calcium entry blockers in anesthetized dogs. Hypertension 5:482–488
6. Leonetti G, Cuspidi C, Sampieri L, Terzoli L, Zanchetti A (1982) Comparison of cardiovascular, renal and humoral effects of acute administration of two calcium channel blockers in normotensive and hypertensive patients. J Cardiovasc Pharmacol 4 (suppl 3):319–324
7. Leonetti G, Gradnik R, Terzoli L, Fruscio M, Rupoli L, Cuspidi C, Sampieri L, Zanchetti A (1986) Effects of single and repeated doses of the calcium antagonist felodipine on blood pressure, renal function, electrolytes and water balance and renin-angiotensin-aldosterone system in hypertensive patients. J Cardiovasc Pharmacol 8:1243–1248
8. Leonetti G, Gradnik R, Terzoli L, Fruscio M, Rupoli L, Zanchetti A (1984) Felodipine, a new vasodilating drug: blood pressure, cardiac, renal and humoral effects in hypertensive patients. J Cardiovasc Pharmacol 6:392–398
9. Leonetti G, Sala C, Bianchini C, Terzoli L, Zanchetti A (1980) Antihypertensive and renal effects of orally administered verapamil. Eur J Clin Pharmacol 18:375–382
10. Di Bona GF, Saxin LL (1984) Renal tubular site of action of felodipine. J Pharmacol Exp Ther 228:420–424

Subject Index